Understanding the Molecular Crosstalk in Biological Processes

Edited by Mohamed A. El-Esawi

Published in London, United Kingdom

IntechOpen

Supporting open minds since 2005

Understanding the Molecular Crosstalk in Biological Processes
http://dx.doi.org/10.5772/intechopen.77821
Edited by Mohamed A. El-Esawi

Contributors
Ioana Mozos, Dana I Stoian, Zsuzsanna Suba, Nicola Di Daniele, Annalisa Noce, Oksana Rekovets, Yuriy Sirenko, Olena Torbas, Mohamed A. El-Esawi

Notice
Statements and opinions expressed in the chapters are these of the individual contributors and not necessarily those of the editors or publisher. No responsibility is accepted for the accuracy of information contained in the published chapters. The publisher assumes no responsibility for any damage or injury to persons or property arising out of the use of any materials, instructions, methods or ideas contained in the book.

First published in London, United Kingdom, 2020 by IntechOpen
IntechOpen is the global imprint of INTECHOPEN LIMITED, registered in England and Wales, registration number: 11086078, 7th floor, 10 Lower Thames Street, London,
EC3R 6AF, United Kingdom
Printed in Croatia

British Library Cataloguing-in-Publication Data
A catalogue record for this book is available from the British Library

Additional hard and PDF copies can be obtained from orders@intechopen.com

Understanding the Molecular Crosstalk in Biological Processes
Edited by Mohamed A. El-Esawi
p. cm.
Print ISBN 978-1-78984-892-2
Online ISBN 978-1-78984-893-9
eBook (PDF) ISBN 978-1-83881-965-1

We are IntechOpen,
the world's leading publisher of
Open Access books
Built by scientists, for scientists

4,900+
Open access books available

123,000+
International authors and editors

140M+
Downloads

Our authors are among the

151
Countries delivered to

Top 1%
most cited scientists

12.2%
Contributors from top 500 universities

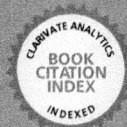

Interested in publishing with us?
Contact book.department@intechopen.com

Numbers displayed above are based on latest data collected.
For more information visit www.intechopen.com

Meet the editor

Dr. Mohamed Ahmed El-Esawi is a visiting research fellow at the University of Cambridge in the United Kingdom, and an associate professor of molecular genetics at the Botany Department of Tanta University in Egypt. Dr. El-Esawi received his BSc and MSc from Tanta University, and his PhD degree in Plant Genetics and Molecular Biology from Dublin Institute of Technology, Technological University Dublin, in Ireland. Afterwards, Dr. El-Esawi joined the University of Warwick in the United Kingdom, University of Sorbonne (Paris VI) in France, and University of Leuven (KU Leuven) in Belgium as a visiting research fellow. His research focuses on plant genetics, genomics, molecular biology, molecular physiology, developmental biology, plant–microbe interaction, and bioinformatics. He has authored several international journal articles and book chapters, and participated at more than 60 conferences and workshops worldwide. Dr. El-Esawi has received several awards and is currently involved in several research projects on biological sciences.

Contents

Preface

Understanding the aspects related to crosstalk in complex biological processes in different organisms, such as plants, microorganisms, animals, and humans, is of utmost importance. This provides the information needed to achieve biological advancements. This book highlights the advances made in this field, which cover crosstalk in different biological processes. Current research trends, research directions, and challenges are also addressed. This book will provoke the interest of various readers, researchers, and scientists, who will find this information useful for the advancement of their research on biological processes in living organisms.

The book's five chapters present an introduction to the crosstalk platform in understanding the biological processes in different living organisms. They also discuss the crossroad between obesity and cancer disease. Furthermore, they review the main mechanisms linking oral and cardiovascular disorders, the pathologies that could be linked, and the possibilities for prophylactic and therapeutic interventions.

The book editor would like to thank Ms. Romina Skomersic and Mr. Gordan Tot, Publishing Process Managers, for their wholehearted cooperation in the publication of this book.

Mohamed Ahmed El-Esawi, PhD
Sainsbury Laboratory,
University of Cambridge,
Cambridge, United Kingdom

Botany Department, Faculty of Science,
Tanta University,
Egypt

Chapter 1

Introductory Chapter: Crosstalk Approach for a Deeper Understanding of the Biological Processes

Mohamed A. El-Esawi

1. Introduction

Molecular signaling has been widely studied in the recent years in order to investigate the different biological processes in living organisms. Information provided by this approach has been utilized to unravel the various functions of different molecules or organs in cells, which in turn facilitate the understanding of the molecular mechanisms underlying the physiological and biochemical processes in these organisms. Therefore, nowadays it is much easier to understand how the biological processes are regulated and controlled inside the organism cells. Understanding the crosstalk and molecular signaling pathways could also help to understand the gene regulatory networking. In plants, studying the signaling processes and crosstalk at the physiological, biochemical, and molecular levels would definitely help to improve the plant growth, development, survival, and productivity as well as to adapt plant crops to the challenging environmental conditions including abiotic and biotic stresses [1–5]. Furthermore, in animals and human, revealing the crosstalk in the biological processes leads to understanding how diseases can be controlled and treated. Therefore, more studies should discuss this matter at the different levels within living organisms. Such kind of information will definitely help to develop different advanced strategies to understand and control the cellular biological processes at different levels.

2. Crosstalk in biological processes

Several earlier studies have reported the crosstalk approach in understanding biological processes in different living organisms. For example, in plants, El-Esawi et al. [1] revealed that Trp triad substitution mutants at W400F and W324F positions can be photoreduced in whole cell extracts, albeit with reduced efficiency. The flavin signaling state (FADH°) has been shown to be stabilized in an in vivo context. These results confirmed that in vivo modulation by metabolites in the cellular environment could has a key role in cryptochrome signaling and is discussed with regard to the possible impacts on the conformation of the C-terminal domain to create the biologically active conformational state. Furthermore, El-Esawi et al. [2] addressed that the blue-light induced biosynthesis of reactive oxygen species may contribute to the signaling mechanism of *Arabidopsis* cryptochrome. El-Esawi et al. [3] also addressed the processes of micropropagation technology and its applications in crop improvement. Nonzygotic embryogenesis and somatic hybridization processes have been explained and assisted in plant development and crop improvement [4, 5]. Moreover, studying the physiological, biochemical, and molecular processes in plants helped to understand the plant development and to develop improved crop varieties tolerant to different environmental stresses [5–15]. In addition, earlier studies reported the importance of crosstalk in understanding the biological processes in other living organisms such as animals and humans. These approaches and processes could be discussed for further understanding and improvement.

Author details

Mohamed A. El-Esawi
Botany Department, Faculty of Science, Tanta University, Tanta, Egypt

*Address all correspondence to: mohamed.elesawi@science.tanta.edu.eg

IntechOpen

References

[1] El-Esawi M, Glascoe A, Engle D, Ritz T, Link J, Ahmad M. Cellular metabolites modulate *in vivo* signaling of *Arabidopsis* cryptochrome-1. Plant Signaling & Behavior. 2015;**10**:e1063758

[2] El-Esawi M, Arthaut L, Jourdan N, d'Harlingue A, Martino C, Ahmad M. Blue-light induced biosynthesis of ROS contributes to the signaling mechanism of *Arabidopsis* cryptochrome. Scientific Reports. 2017;**7**:13875

[3] El-Esawi MA. Micropropagation technology and its applications for crop improvement. In: Anis M, Ahmad N, editors. Plant Tissue Culture: Propagation, Conservation and Crop Improvement. Singapore: Springer; 2016. pp. 523-545

[4] El-Esawi MA. Nonzygotic embryogenesis for plant development. In: Anis M, Ahmad N, editors. Plant Tissue Culture: Propagation, Conservation and Crop Improvement. Singapore: Springer; 2016. pp. 583-598

[5] El-Esawi MA. Somatic hybridization and microspore culture in *Brassica* improvement. In: Anis M, Ahmad N, editors. Plant Tissue Culture: Propagation, Conservation and Crop Improvement. Singapore: Springer; 2016. pp. 599-609

[6] El-Esawi MA, Alayafi AA. Overexpression of rice *Rab7* gene improves drought and heat tolerance and increases grain yield in rice (*Oryza sativa* L.). Genes. 2019;**10**:56

[7] El-Esawi MA, Al-Ghamdi AA, Ali HM, Alayafi AA, Witczak J, Ahmad M. Analysis of genetic variation and enhancement of salt tolerance in French pea (*Pisum sativum* L.). International Journal of Molecular Sciences. 2018;**19**:2433

[8] El-Esawi MA, Alaraidh IA, Alsahli AA, Ali HM, Alayafi AA, Witczak J, et al. Genetic variation and alleviation of salinity stress in barley (*Hordeum vulgare* L.). Molecules. 2018;**23**:2488

[9] El-Esawi MA, Alaraidh IA, Alsahli AA, Alamri SA, Ali HM, Alayafi AA. *Bacillus firmus* (SW5) augments salt tolerance in soybean (*Glycine max* L.) by modulating root system architecture, antioxidant defense systems and stress-responsive genes expression. Plant Physiology and Biochemistry. 2018;**132**:375-384

[10] El-Esawi MA, Alaraidh IA, Alsahli AA, Alzahrani SM, Ali HM, Alayafi AA, et al. *Serratia liquefaciens* KM4 improves salt stress tolerance in maize by regulating redox potential, ion homeostasis, leaf gas exchange and stress-related gene expression. International Journal of Molecular Sciences. 2018;**19**:3310

[11] El-Esawi MA, Al-Ghamdi AA, Ali HM, Alayafi AA. *Azospirillum lipoferum* FK1 confers improved salt tolerance in chickpea (*Cicer arietinum* L.) by modulating osmolytes, antioxidant machinery and stress-related genes expression. Environmental and Experimental Botany. 2019;**159**:55-65

[12] Jourdan N, Martino C, El-Esawi M, Witczak J, Bouchet P-E, d'Harlingue A, et al. Blue light dependent ROS formation by *Arabidopsis* Cryptochrome-2 may contribute towards its signaling role. Plant Signaling & Behavior. 2015;**10**:e1042647

[13] Vwioko E, Adinkwu O, El-Esawi MA. Comparative physiological, biochemical and genetic responses to prolonged waterlogging stress in okra and maize given exogenous ethylene priming. Frontiers in Physiology. 2017;**8**:632

[14] El-Esawi MA, Al-Ghamdi AA, Ali HM, Ahmad M. Overexpression of *AtWRKY30* transcription factor enhances heat and drought stress tolerance in wheat (*Triticum aestivum* L.). Genes. 2019;**10**(2):163

[15] El-Esawi MA, Alayafi AA. Overexpression of *StDREB2* transcription factor enhances drought stress tolerance in cotton (*Gossypium barbadense* L.). Genes. 2019;**10**:142

Crossroad between Obesity and Cancer: A Defective Signaling Function of Heavily Lipid-Laden Adipocytes

Zsuzsanna Suba

Abstract

Obesity and its comorbidities exhibit a gender-related dimorphism. Obese males tend to accrue more visceral fat leading to abdominal adiposity, which shows a strong correlation with serious obesity-associated comorbidities, cardiovascular diseases and cancers. In contrast, obese females accumulate excessive fatty tissue predominantly subcutaneously enjoying strong protection from the obesity-related diseases. The health advantage of obese women as compared with obese men may be attributed to their higher estrogen production and an increased transactivation of estrogen receptors (ERs). The recently clarified intracrine, paracrine, and endocrine functions of adipose tissue illuminate that concentrations of estrogens and the suitable expression and activity of ERs strongly define all regulatory functions in both men and women. All well-known cancer risk factors are in correlation with defects of estrogen signaling in partnership with glucose intolerance as estrogen regulates all steps of glucose uptake. In central obesity, increased secretions of cytokines and growth factors are not causal factors of developing insulin resistance, and unrestrained cell proliferation, but rather, they are compensatory processes so as to increase estrogen synthesis and ER transactivation. In conclusion, a causal therapy against obesity and obesity-related diseases aims to improve estrogen signaling in both men and women.

Keywords: obesity, insulin resistance, estrogen signaling, cancer risk, cardiovascular disease, estrogen receptor, IGF-1, low-grade inflammation, inflammatory cytokines

1. Introduction

Gender-related differences in the risk for obesity-associated serious diseases, such as type 2 diabetes, cardiovascular lesions, and cancers, are well established [1, 2].

Fat deposition exhibits a strong sexual dimorphism defined by genetic factors [3]. Among healthy lean people, women have a higher body fat content than men. In obesity, males tend to accrue excessive visceral fat leading to abdominal adiposity, which shows a strong correlation with serious obesity-associated comorbidities. In contrast, obese females accumulate excessive fat deposition predominantly subcutaneously in the gluteofemoral region, gaining protection from the obesity-related diseases. However, after menopause, excessive fat depositions in obese women show a shift to favor the visceral depot in a male-like manner, resulting in an increased risk for obesity and associated comorbidities [4]. Considering the gender and age-related differences in the epidemiology of obesity and obesity-associated diseases, an imbalance of sexual hormone signaling emerges as a crucial player in the dysregulation of adipose tissue in obese patients [5].

Both clinical and experimental results suggest that obesity, particularly a visceral accumulation of fatty tissue, leads to insulin resistance, which is a defect of insulin-assisted cellular glucose uptake [6]. Dysregulation of glucose uptake may induce serious disorders in the gene regulation of cellular metabolism, growth, differentiation, and mitotic activity as glucose is the most important fuel of genomic machinery in mammalian cells [7].

Insulin resistance was the first one among revealed cancer risk factors detrimentally affecting cellular signaling functions [8]. Later on, defective estrogen signaling emerged as a second cancer risk factor, which directly deteriorates the functions of genomic machinery [9]. In the meantime, a central role of sexual steroid synthesis and estrogen signaling emerged as the chief regulator of all extragonadal tissues including adipose tissue [10]. Finally, a fundamental role of estrogen signaling in the regulation of both somatic and reproductive functions of mammalian cells was exposed [11].

The recently clarified intracrine, paracrine, and endocrine functions of adipose tissue illuminate that concentrations of estrogens, and the appropriate expression and activity of their receptors, strongly define the regulatory functions of central adipose tissue in both men and women [12]. Decreased estrogen signaling induces central obesity through an increased lipogenesis and insulin resistance of abdominal adipocytes. The excessive lipid storage strongly inhibits the signaling crosstalk between adipose cells and visceral organs, including intestine, liver, pancreas, kidneys and cardiovascular system. The disturbed inter-tissue crosstalk may cause defects in both metabolic and mitotic activities of parenchymal and stromal cells of different organs and induces serious, life-threatening diseases.

Understanding the pivotal regulatory roles of healthy abdominal fatty tissue in signal transduction may appropriately answer the question: how can abundant visceral fat deposition lead to serious, life-threatening diseases, such as cardiovascular lesions and malignancies in different organs?

2. Weak estrogen signaling in men and postmenopausal women provokes an increased prevalence of abdominal obesity and glucose intolerance

In obese patients, an excessive fat deposition in the intra-abdominal adipose depot is strongly associated with an increased risk for type 2 diabetes, cardiovascular diseases, and cancers developing at different sites [13]. Gender-related dimorphism of adipocyte behavior and fatty tissue accumulation suggests that health advantages of obese women as compared with obese men may be attributed to stronger estrogen signaling and an increased activation of estrogen-regulated genes [14].

In obese men, an excessive abdominal fatty tissue deposition is characteristic. Male-like abdominal obesity is strongly associated with different stages of insulin resistance and an increased risk for obesity-associated comorbidities including cancer [3].

Among obese premenopausal women with regular ovulatory cycles, a female-type gluteofemoral deposition of adipose tissue may be associated with the maintenance of healthy insulin sensitivity, whereas they have no increased risk for either cardiovascular or neoplastic diseases [15]. Abdominal adipocytes of healthy women exhibit higher insulin sensitivity as compared with those of men [16]. In contrast, in premenopausal obese women with anovulatory infertility or disorders of menstrual cycles, a male-like abdominal obesity and deepening insulin resistance may develop in correlation with the defect of estrogen signaling [17].

In women, menopause is a physiological model revealing how decreased estrogen level may induce adiposity and its comorbidities. In obese, postmenopausal women, an increased inclination to a male-like central obesity and associated insulin resistance may be experienced [4, 5]. Increased prevalence of abdominal obesity and obesity-related diseases in postmenopausal women strongly suggests that decreased estrogen signaling may have pivotal roles in the dysregulation of lipid metabolism, glucose tolerance, and cell proliferation [18, 19]. In contrast, hormone replacement therapy after menopause reduces the accumulation of visceral adipose tissue and improves insulin sensitivity via an activation of estrogen-regulated genes [20].

Inherited serious mutations on estrogen receptor alpha gene (ESR1) or aromatase coding CYP19 gene may result in extreme defects of estrogen signaling and lead to insulin-resistant states and premature cardiovascular diseases in both male and female cases [21–23].

In rodents, estrogen withdrawal via ovariectomy consistently increases adiposity-associated body weight and glucose intolerance, while estrogen treatment results in weight loss and a restoration of insulin sensitivity [24, 25]. Aromatase knock-out (ArKO) transgenic mice with inactivated aromatase enzyme are unable to synthesize estrogen and exhibit increasing obesity and insulin resistance with associated compensatory hyperinsulinemia in males and females [26]. ER-alpha knock-out (ERαKO) male and female mice similarly exhibit metabolic syndrome-like phenotypes, including obesity, glucose intolerance, hyper-insulinemia, and decreased energy expenditure [27].

In ovariectomized obese mice, inoculated T47D tumor cells exhibited an intense proliferation, while estrogen supplementation suppressed the survival of inoculated tumor cells [28]. Ovariectomized female mice were fed diets to induce obesity and were inoculated with mouse mammary cancer cells. Estrogen substitution triggered a loss of body fat, improved insulin sensitivity, and suppressed the proliferation of inoculated tumors [29].

In conclusion, either estrogen deficiency or estrogen receptor (ER) resistance may have a crucial role in the dysregulation of both lipid storage and glucose uptake in adipocytes resulting in obesity and type 2 diabetes as clinical manifestations. Estrogen substitution seems to be a good strategy against obesity and associated chronic diseases including cancer.

3. Obesity-associated failure of estrogen signaling is a stronger cancer risk for males having low baseline estrogen levels

Systematic review and meta-analysis of literary data help to assess the associations between obesity and cancer risk in males and females [30]. Among men, obesity disproportionately increases the prevalence of cancers developing at different sites. These observations strongly suggest that the health advantage in obese women may be attributed to their estrogen predominance supplying defense against failures in DNA replication [31].

In epidemiologic studies, a male predominance was experienced in *esophageal adenocarcinoma* incidence [32]. In a clinical study, *oral cancer* showed a conspicuous male predominance, while from 50 to 52 years of age, oral cancer incidence among postmenopausal women showed a slow and, later, a steep increase in correlation with the loss of estrogen exposure [33]. Results of meta-analyses do indicate that the association between obesity and an increased risk for *gastric cancer* is stronger in men than in women and the correlation significantly increases with increasing BMI among men [34]. *Colorectal cancer* shows a high prevalence among men, while women are relatively protected against this tumor [35]. Furthermore, hormone replacement therapy provides an additive protective effect against colon tumors in postmenopausal women. Obesity increases the risk for colorectal cancer in both men and women; while in males, associations between excessive adiposity and colon cancer risk seem to be much stronger [30]. Obesity and related metabolic disorders are high risk factors for *primary liver cancer* [36]. The associations are stronger in men, especially in patients with underlying liver disease, such as HCV infection or cirrhosis.

Unexpectedly, the risk for *kidney cancer* is higher in obese women than in men with increasing BMI [37]. Overweight and obese patients were found to have elevated risks of renal cell cancer in a dose-response manner, with an estimated 24% increase for men and a 34% increase for women by every 5 kg/m^2 increase in body mass index (BMI) [30]. Conversely, data from the National Cancer Database during a 10-year period were analyzed and a ratio of 1.65 of renal cell carcinoma risk for males compared to females was established disregarding the impacts of differences in BMI. Nevertheless, the mean age of kidney cancer cases was greater in females (64.3) than in males (60.9) (p < 0.001) [38]. In obese women, a long-lasting postmenopausal decrease in estrogen synthesis may be an additional risk for kidney cancer, provoking an inverse gender-related disparity for kidney cancer in an aged population.

Breast cancer in women is the most frequently diagnosed malignant tumor, while in men, it is an extremely rare disease, accounting for less than 1% of all breast cancer cases [39]. Since high estrogen levels are mistakenly regarded as fuels for breast cancer growth, the strong protection of males against breast cancer may be deceivingly attributed to their physiologically lower estrogen levels. The causal factors of male breast cancer are quite similar to those of female breast malignancies—obesity, type 2 diabetes and male infertility [39], which are strongly associated with insulin resistance and deficient estrogen signaling. Healthy female breasts exhibit high estrogen demand attributed to their physiologic cycling activity, and they are highly vulnerable even to slightly defective estrogen signaling. In contrast, in resting male breasts, very serious defects of estrogen signaling may be regarded as dangerous regulatory disorders leading to an increased risk for cancer [31].

Obesity equivocally increases the incidence of breast cancer among males, presumably through similar mechanisms as in case of females [40]. However, among women, an apparently ambiguous interaction can be observed between obesity and breast cancer risk depending on the menopausal status of patients [41]. In young women before menopause, obesity is erroneously regarded to exhibit a protective effect against breast cancer. Conversely, in postmenopausal older cases, there is a strong direct correlation between adiposity and mammary cancer risk. Obesity is associated with dysmetabolism and endangers the healthy equilibrium of sexual hormone production. In reality, among premenopausal women, obesity-associated insulin resistance is moderate, being counteracted by their maintained or increased circulating estrogen levels [15, 42]. In conclusion, it is not obesity but rather the still suitable estrogen level that may be protective against breast cancer in young women. Obese postmenopausal women, who had never used hormone replacement therapy (HRT), exhibit fairly high breast cancer risk in association with their continuously decreasing estrogen levels [43]. In contrast, obese postmenopausal women using hormone replacement therapy (HRT) show a significantly reduced breast cancer risk attributed to the protective effect of estrogen substitution [44, 45].

The higher risk for obesity-associated cancer among men and postmenopausal women as compared with premenopausal cases suggests that the weaker the estrogen signaling in abdominal adipocytes, the stronger is the defect of DNA repair leading to an increased risk for cancer [15, 42].

4. Central adipose tissue is a hub in the signaling network that controls and regulates visceral organs via an inter-tissue crosstalk

Adipose tissue stores excess calories in the form of lipid. In addition, adipose tissue is an endocrine organ regulating the functional activity of both adjacent and remote organs [46].

Adipose tissue is principally deposited in two locations. *Centrally positioned* fatty tissue within the trunk and abdomen closely surrounds the visceral organs, while the *subcutaneously positioned* adipose tissue covers the skeletal muscles [2]. The close vicinity between central fatty tissue and internal organs suggests that adipocyte signaling in this region may regulate vital functions affecting the health of whole body, while subcutaneous adipocytes primarily confer their messages to skeletal muscles and skin. Differences in fatty tissue locations strongly suggest that in pathological situations included obesity, the dysregulation of central adipocytes is a much higher health risk as compared with that of their peripheral counterparts.

In healthy nonobese people, abdominal adipose tissue provides a strong mechanical support for all internal organs [2]. Visceral fat is largely located in the omental and mesenteric adipose tissue in the vicinity of stomach, intestines, liver, and pancreas. Adipose tissue deposition is also characteristic within the visceral pericardium surrounding the myocardium and coronary arteries. Kidneys and the attached adrenal glands are embedded into an abundant fatty tissue capsule. Central adipose tissue is a crucial fat store providing standby energy for physiologic functions and even for the functional activation of visceral organs among various circumstances.

On the other hand, central adipose tissue is a complex and highly active endocrine organ exerting essential regulatory functions through autocrine, paracrine, and endocrine mechanisms [47]. In addition to the mass of adipocytes, adipose tissue contains a connective tissue matrix comprising fibrocytes, nerves, vessels, and immune cells functioning as an integrated unit.

The receptor systems of adipocytes, fibrocytes, and immune cells collect numerous afferent signals arriving from adjacent or remote organs and the central nervous system, while they respond through the secretion of signaling molecules, such as steroid hormones, adipokines, growth factors, and cytokines.

5. Circulating and locally synthesized estrogen hormones regulate the functional activity of adipose tissue

Sex steroids, particularly estrogens, play a pivotal role in the regulation of all tissues in the body [10]. Estrogen-activated ERs are central regulators of the metabolic status, growth, differentiation, and proliferation of cells in mammalians. Estrogen-regulated genes control signaling functions of the whole body via a balanced activation of ERs through liganded and unliganded pathways [48]. Estrogens are the regulators of crosstalks between adipose tissue and adjacent visceral organs via an appropriate expression of signaling molecules in both males and females.

Physiological cellular mechanisms require not only suitable estrogen concentrations but also appropriate expression and activity of ERs in the targeted tissues.

The presence of both isomers, ER-alpha and ER-beta, was confirmed in adipocytes deriving from both subcutaneous and intra-abdominal adipose tissue, suggesting that adipose tissue function is strongly defined by suitable estrogen signaling [49]. Lower prevalence of visceral adiposity in women than in men arises from the high expression and activity of ERs in female abdominal adipocytes being capable of downregulating both adipogenesis and lipid storage [50].

The gonads, ovary, and testis are the primary sites of estrogen synthesis in mammals. Extragonadal estrogen synthesis in adipose tissue was first published in 1974, based on the unexpected observation that androgens were converted to estrogens by aromatase enzyme in adipose tissue [51]. From that time onward, estrogen synthesis and the expression of estrogen receptors have been revealed in several organs [52]. All tissues expressing estrogen receptors are considered to be targets of estrogenic regulation.

Adipose tissue is considered to be a major source of estrogen synthesis among extragonadal sites in both women and men, and it increasingly contributes to the circulating estrogen concentration with aging [12]. In adipose tissue, C19 steroids are essential precursors of estrogen synthesis and the locally synthesized CYP19 aromatase enzyme is able to convert C19 steroids to estrogens [53]. The prerequisite of estrogen synthesis is the local expression of aromatase enzyme, which strongly defines the local estrogen production. In extra-gonadal tissues, the expression level of aromatase enzyme shows strong parallelism with the intensity of estrogen synthesis. Extragonadal sites included adipose tissue are unable to synthesize C19 steroids; hence, their estrogen synthesis is limited by the precursor supply from external sources [53].

Estrogens synthesized in adipose tissue are thought to act locally, in an *autocrine manner*, while they may have sufficient concentration in the adjacent organs too so as to induce ER activation in a *paracrine manner*. The systemic, *endocrine effects* of locally synthesized estrogens are limited [54]. Appropriate local estrogen concentration may exert its biological activity via activation of ERs, which are members of the nuclear receptor superfamily.

Increased estrogen concentrations upregulate estrogen signaling through a DNA stabilizer circuit and lead to a higher expression of ERs, genome stabilizer proteins, and aromatase enzyme [11, 55]. In contrast, estrogen deficiency or a defect of ER activation causes dysregulation in adipocytes and leads to a derangement of their signaling functions.

The noteworthy volume of central adipose tissue and the remarkable estrogen synthesis of adipocytes may support the fact that estrogen signaling has a crucial role in controlling and orchestrating the signaling network of both adjacent organs and the whole body.

6. Estrogen-regulated genes orchestrate the physiological functions of adipose tissue and visceral organs

Estrogen-activated ERs are the chief regulators of somatic and reproductive cellular functions suggesting that a defect in estrogen signaling may produce dysregulation and leads to serious diseases [48]. Current literary data support that activated ERs in adipocytes protect against fat deposition, insulin resistance, inflammation, and fibrosis of adipose tissue [Davis]. Moreover, body fat mass deposition is defined by the level of estrogen signaling in both men and women [56]. Healthy, insulin-sensitive adipose tissue regulates the glucose homeostasis and balanced lipolysis/lipogenesis of adjacent tissues, included liver, skeletal muscles, and further organs via a tissue crosstalk [57].

6.1 Adipose tissue

Estradiol-induced activation of ER-alpha upregulates all steps of cellular glucose uptake increasing the *insulin sensitivity* of adipocytes [58]. In mature adipocytes, estradiol treatment enhances insulin-assisted glucose uptake through liganded and unliganded activations of ER-alpha. Estradiol is capable of stimulating an increased tyrosine phosphorylation of insulin receptor substrate-1 [59]. In vitro, estradiol activates adenosine monophosphate-activated protein kinase (AMPK) and protein kinase B (AKT) inducing unliganded ER signaling even in the absence of insulin [60]. In human adipocytes, GLUT-4 abundance shows high correlation with insulin responsiveness. In 3T3-L1 adipocytes, estradiol treatment facilitated glucose uptake via an increased expression and intracellular translocation of glucose transporters (GLUTs) [61].

In obesity, GLUT-4 content showed a 40% decrease in adipocyte membranes, while in estrogen-deficient adipocytes deriving from either lean or obese PCOS cases, a 36% decrease of GLUT-4 content was experienced [62]. In animal experiments, female, ovariectomized high-fat-fed mice exhibited obesity, and decreased levels of glucose transporter 4 (GLUT4) and ER-alpha protein were found in their abundant visceral fatty tissue coupled with increasing insulin resistance [63].

In women, a higher ER expression in abdominal adipocytes may be associated with higher insulin sensitivity of fatty tissue and a lower susceptibility to inflammation and fibrosis compared to men [56]. Estrogen receptor α gene expression levels are reduced in the adipose tissue of obese women compared to those found in normal weight females [64].

The identification of a mesenteric estrogen-dependent adipose gene (MEDA-4) revealed that estrogen signaling may selectively regulate mesenteric adipocytes in a depot-specific manner [65]. In ovariectomized female mice, an increased MEDA-4 expression in mesenteric adipose tissue was capable of increasing adipose tissue expansion, while estradiol substitution reduced MEDA-4 expression and normalized the balance of lipolysis and lipogenesis in central adipocytes.

In conclusion, estrogen-regulated genes play crucial roles in both lipogenic control and glucose uptake of adipocytes. In contrast, estrogen deficiency-associated insulin resistance and lipogenesis further deteriorate estrogen signaling and strongly decrease the expression of estrogen-regulated genes via a vicious circle [66].

6.2 Liver

In the liver, estrogen-regulated genes control the synthesis of various proteins. Estrogen signaling regulates lipoprotein synthesis, lipogenesis, and lipolysis in hepatocytes. Estradiol controls the synthesis of blood clotting factors, including factors II, VII, IX, X, and plasminogen [67]. Estrogen regulates glucose uptake and glucose homeostasis in hepatocytes, improving insulin sensitivity [68]. In the pathologic conditions of the liver, estradiol treatment exerts anti-inflammatory and antineoplastic impacts through ER-alpha and, predominantly, ER-beta activation [69]. In contrast, a longer duration of estrogen deficiency increases the risk of hepatic fibrosis among postmenopausal women with nonalcoholic fatty liver disease [70]. In female mice, estrogen treatment prevents the development of insulin resistance and low-grade inflammation in adipose tissue [71, 72]. Recent literary data strongly confirm that estrogen hormone protects against the development and progression of hepatocellular carcinoma (HCC) [73]. Estrogen treatment also mediates antitumor effects in intrahepatic cholangiocarcinoma cases via a balance of ER-alpha and ER-beta-regulated pathways [74]. Bilateral oophorectomy in premenopausal women increases the risk of hepatocellular carcinoma via estrogen withdrawal [75].

6.3 Pancreas

In pancreatic islands, the estradiol activation of ER-alpha regulates β cell proliferation during pancreatic development [76]. Estradiol activates the expression of insulin gene and increases insulin synthesis in the β cells [77, 78]. Estradiol activation of ERs inhibits apoptotic death of β cells in case of inflammatory insult [77]. Estradiol promotes β cell recovery after pancreatic injury [80]. In rodent models of type 2 diabetes, estrogen treatment reduced lipid synthesis and prevented β cell failure in pancreatic islets [79]. Correlations between parity and pancreatic cancer risk were studied in a meta-analysis of epidemiologic studies and the findings suggest that higher parity is associated with a decreased risk of pancreatic cancer [80].

6.4 Gut

Local estrogen synthesis and ER expression in the gut are important players in intestinal homeostasis. Estrogen signaling regulates the integrity and function of intestinal epithelium controls intestinal epithelial barriers and reduces intestinal permeability [81]. Estrogen signal provides protection against duodenal ulcer, inflammatory bowel disease, and colon cancer in both animal experiments and clinical studies [82]. Overexpression of ER-beta in healthy human colon coupled with its reduced expression in colon cancer suggests that the activation of ERs, particularly ER-beta, might be particularly involved in the estrogen-mediated protection against colon tumors [35].

The microbes that colonize the human gut (microbiome) may play contributory roles in health and disease. Intestinal microbiome participates in polysaccharide breakdown, nutrient absorption, inflammatory responses, and bile acid metabolism [83]. Gut microbiome is one of the principal regulators of circulating estrogen levels [84]. Microbes in the gut regulate estrogen levels through secretion of β-glucuronidase, an enzyme that converts conjugated estrogens into their biologically active forms. Lower microbial density or a dysbiosis of gut microbiome leads to a decrease in estrogen deconjugation and results in a reduction of both luminal and circulating free estrogen levels.

The low level of luminal estrogens induces gut diseases included inflammation and cancer, while an associated decrease in circulating estrogens may contribute to the development of systemic diseases including obesity and type 2 diabetes.

In medical practice, a fecal microbiome transplant (FMT) from healthy individuals to ill obese patients seems to be a promising way to improve the gut microbiome and for the treatment of obesity and associated metabolic syndrome [85]. Presumably, FMT may provide estrogen hormones and estrogen receptors for recipient patients, which beneficially restore intestinal estrogen signaling besides the re-colonization of microbes capable of increasing free estrogen levels.

In conclusion, alterations in the density and composition of gut microbiome deteriorate the metabolic profile of omental and mesenteric adipose tissues via deficient estrogen signaling and lead to the development of abdominal obesity and metabolic syndrome.

6.5 Kidney

Estrogens have nephroprotective effects. The most important actions of estrogen hormones are represented by the protective effects attenuating glomerulosclerosis and tubulointerstitial fibrosis. Estrogens exert favorable effects on renal osteodystrophy via an improvement of phosphorus-calcium transport equilibrium in renal tubules [86]. Recently, a meta-analytic study supported that oral contraceptive use may reduce the risk of kidney cancer in women, especially for long-term users [87].

6.6 Cardiovascular system

Premenopausal women have a much lower incidence and prevalence of cardio-vascular diseases (CVDs) compared to men of the same age [88]. This sex difference in favor of women suddenly disappears after menopause, suggesting that reduced levels of ovarian hormones constitute a major risk factor for the development of CVD in postmenopausal women.

In central obesity, the associated insulin resistance results in a compensatory high insulin and IGF-1 synthesis stimulating ovarian androgen production at the expense of reduced estrogen synthesis [89]. Hyperinsulinemia promotes adrenal androgen synthesis as well by means of an increased adrenal sensitivity to adreno-corticotropin [90, 91]. Excessive androgen synthesis would provide precursors for an increased estrogen production; however, in insulin-resistant patients, gonads exhibit a decreased capacity to convert androgen to estrogen attributed to a reduc-tion of aromatase enzyme activity [92].

Estrogens are cardioprotective hormones having crucial role in the maintenance of physiologic serum lipid levels. Postmenopausal hormone replacement therapy may reduce the risk of cardiovascular diseases via lowering total cholesterol, LDL cholesterol, and triglyceride levels [93]. Estrogens show antihypertensive activity as well. It is capable of downregulating the components of the renin-angiotensin system (RAS) [94]. Estradiol inhibits the excessive synthesis of vasoconstrictor endothelin and improves endothelial dysfunction [95]. In animal experiments, estrogen receptor-alpha mediated the protective effects of estrogen in response to vascular injury [96].

The risk of mortality for ischemic heart disease is highly increased in patients with left-sided breast cancer as they receive a higher dose of radiation therapy affecting the heart than patients with right-sided tumors [97]. This correlation may be explained by the destructive effect of radiation on the estrogen-synthesizing epicardial fatpad.

7. Secretory functions of abdominal adipose tissue in healthy lean and obese patients

Abdominal adipose tissue has crucial physiological secretory functions [47]. Sex steroids, adipokines, cytokines, and growth factors are important signaling molecules in adipose tissue and their well-regulated activation ensures the health of the whole body.

In central obesity, extreme fat accumulation and a concomitant insulin resis-tance of abdominal adipose tissue deteriorate all regulatory functions of adipocytes. Dysregulation of adipocyte signaling induces an alarm reaction in all adjacent visceral organs and they inform adipocyte ERs about their functional difficulties via signaling molecules. ERs in adipocytes recognize the emergency situation and may initiate the restoration of ER signaling through increased estrogen synthesis and enhanced liganded and unliganded activations of ERs [98].

7.1 Sexual steroids

In adipose tissue, suitable estrogen signaling regulates the expression of numerous genes and the harmonized synthesis of signaling molecules [67].

In adipocytes, androgen hormones are precursors for the local estrogen synthesis of aromatase enzyme.

7.2 Adipokines

Leptin regulates the energy balance in the hypothalamus exerting anorexino-genic and lipolytic effects. Estrogen treatment increases the expression of leptin receptors amplifying the leptin-sensitivity of various cells [99]. In aromatase knock-out (ARKO) mice, visceral adiposity develops and three times higher levels of circulating leptin were found as compared with control animals [100]. Adiponectin is involved in various inflammatory reactions, modulations of endothelial functions, and is protective against insulin resistance. Oophorectomy increases adiponectin levels in adult mice, while it may be reversed by estradiol substitution [101]. Resistin level increases in parallel with obesity, which may be a compensatory reaction instead of being a causal factor. In subcutaneous adipocytes, an estradiol benzoate injection decreases resistin levels [102].

7.3 Pro-inflammatory cytokines and low-grade inflammation

Pro-inflammatory cytokines are regulatory proteins having great role in the maintenance of genomic stability. In obese patients, strong inflammatory reactions and abundantly expressed signaling molecules occurring in voluminous fatty tissue are mistakenly considered as being causal factors of genomic dysregulation, insulin resistance and its comorbidities [6, 103].

Low-grade inflammation in abundant adipose tissue may be characterized by increased levels of inflammatory cytokines and macrophage and T-cell infiltration [104]. Inflammatory cytokines, including tumor necrosis factor alpha (TNF-α) and interleukin-6 (IL-6), may have not only local effects on adipose tissue but also exert systemic effects on several organs [105]. In adipose tissue, inflammatory cytokines have an important role in the upregulation of estrogen synthesis. Increased pro-inflammatory cytokine level generates an increased expression and activation of aromatase enzyme resulting in higher estrogen concentrations [106].

In ovariectomized mice on high-fat diet, estradiol treatment improved insulin sensitivity and glucose uptake in both adipose tissue and liver and at the same time, decreased the expression of inflammatory cytokines, included TNFα [71]. It was recently found that elevated levels of pro-inflammatory cytokines augment energy expenditure and counteract obesity from animal to human, justifying that pro-inflammatory cytokines have beneficial effects against obesity and obesity-related metabolic disorders [107].

In conclusion, in obesity, deficient estrogen signaling leads to dysregulation of abdominal adipocytes and induces a low-grade inflammatory reaction, while estrogen treatment improves adipocyte signaling function and exerts anti-inflammatory effect.

7.4 Insulin-IGF system

The insulin-like growth factor (IGF) system is involved in regulation and control of physiologic growth and differentiation. Insulin and insulin-like growth factor receptors act as ligand-specific amplitude modulators regulating genes on a common pathway [108]. In obesity, there was a nonlinear relationship of insulin-like growth factor (IGF)-I and IGF-I/IGF-binding protein-3 ratio with indices of adiposity and plasma insulin concentrations [109].

Crosstalks between signaling pathways of ERs and growth factor receptors (IGF-1R, EGFR, VGFR) are well known in both health and disease [110, 111].

Among physiological circumstances, estrogen-activated ERs are capable of either stimulating or silencing cell growth and proliferation, while in tumor cells, an upregulation of ERs is not a proliferative stimulus, but rather, it is an initiator of self-directed death [11].

The IGF system comprises insulin, insulin receptor (IR), IGF-1, IGF-2, each with its receptor (IGF-1R, IGF-2R) as well as the IGF-binding proteins (IGFBPs). Proteins of the IGF system are ubiquitously expressed at different sites included adipose tissue [13]. In the early phases of insulin resistance, an increased IGF-1 level mediates increased insulin synthesis resulting in a compensatory hyperinsulinemia.

Estrogens regulate both insulin-like growth factor 1 (IGF-1) synthesis and the expression of its receptor (IGF-1R) in adipocytes. In turn, an increased synthesis of IGF-1 and its receptor upregulates the AKT and MAPK pathways of growth factor signaling, providing a possibility for an increased unliganded activation of ERs [112]. In an estrogen-deficient milieu, unliganded activation of ERs via IGF-1 receptor signaling pathway exhibits a fundamental role in genome-wide expressions of estrogen-regulated genes [48]. In conclusion, in obesity and insulin resistance, an increased expression and activity of IGF-1 receptors may be a compensatory action against the dysregulation of adipocytes providing a possibility for the restoration of ER signaling.

7.5 Interaction between adipocytes and immune cells

Adipocytes are in signaling crosstalk with immune cells in both healthy and obese adipose tissue. In lean adipose tissue, IL-4 secreted by eosinophil granulocytes and regulatory T (Treg) cells activate M2 type macrophages, which express arginase and anti-inflammatory cytokines such as IL-10. In contrast, in obese adipose tissue, the number of pro-inflammatory immune cells is increased coupled with a decrease in that of anti-inflammatory immune cells [104]. In abundant adipose tissue, neutrophil granulocytes are early responders to inflammatory injuries stimulating both M1 type macrophage infiltration and pro-inflammatory cytokine secretion.

In animal experiments, estrogens are capable of improving metabolic disorders and, at the same time, they exert anti-inflammatory effects [72]. In female mice, estrogen protects from adipocyte hypertrophy and prevents adipose tissue oxidative stress and inflammation. Furthermore, estrogen treatment protected female mice against developing liver steatosis and from becoming insulin resistant.

In conclusion, in obesity, the experienced low-grade inflammation and the increased synthesis of cytokines and growth factors are not causal factors of insulin resistance and cancer development, but rather, they are compensatory efforts so as to restore ER signaling. Estrogen administration seems to be a causal therapy for obesity-related pathologic changes, as the upregulation of ER signaling protects from insulin resistance and decreases inflammatory reactions. Estrogen treatment would provide quite new ways for the prevention and cure of obesity and obesity-related inflammation.

8. Estrogen is protective against cancer

The apparently paradoxical effects of estrogens on both the prevention and promotion of cancers have been debated for over 120 years. However, the pharmaceutical development of endocrine disruptor antiestrogens could not provide appropriate advances in the field of anticancer fight [113].

The blockade of either estrogen-induced (liganded) or unliganded activation of ERs provoked a strong compensatory upregulation on the unaffected domain, while the unbalanced ER activation resulted in modest or even adverse tumor responses and severe toxic symptoms [114]. In contrast, natural estrogens are capable of restoring DNA surveillance even in tumor cells leading to a self-directed death through a balanced liganded and unliganded transactivation of ERs, whereas they may not provoke genomic instability even in sky-high concentrations [115].

9. Conclusion

Changing concepts concerning the etiology of obesity and obesity-related diseases provide certain new strategies against these diseases.

Since estrogen-activated ERs are the chief regulators of the genomic machinery, a weakening or failure of ER signaling attributed to either genetic or environmental factors induces abundant compensatory actions for the restoration of appropriate estrogen signaling. When the restorative processes may compensate the regulatory failures, patients seem to be healthy. In contrast, when the compensatory reactions are insufficient, a deepening regulatory disorder develops and serious diseases may appear as clinical manifestations.

The causal factor of obesity is deficient estrogen signaling in both men and women; however, there is no direct correlation between obesity and cancer risk. All environmental and genetic cancer risk factors, included smoking, sedentary lifestyle, fat-rich diet, light deficiency, anovulatory infertility, and BRCA mutation strongly deteriorate estrogen signaling; however, they may or may not be associated with the development of obesity.

In certain cases, the weakness of estrogen regulation may induce abdominal fat deposition, and at the same time, the deepening insulin resistance of adipose tissue leads to further irreversible defects in expressions of estrogen-regulated genes. In obese patients, metabolic syndrome and type 2 diabetes are well-known risk factors for the two life-threatening disease groups of humans: cardiovascular lesions and malignancies, while the defect of estrogen signaling frequently remains unknown in the background.

In another group of patients, estrogen deficiency or ER resistance develops without obesity. Defective estrogen signaling exhibits a strong direct correlation with insulin resistance as estrogen regulates all steps of cellular glucose uptake in the whole body. Moreover, the deeper the defect of estrogen surveillance, the stronger is the risk for serious chronic diseases included type 2 diabetes, cardiovascular lesions, and cancers. The strong correlation is well known between anovulatory infertility and an increased cancer risk for female organs, while these associations are quite independent of obesity.

In patients showing central obesity, the increased cytokine secretion and the associated low-grade inflammation in their adipose tissue are not causal factors of the development of insulin resistance, but rather, they are compensatory actions for the activation of aromatase synthesis. Similarly, in abundant adipose tissue, the increased synthesis of growth factors included IGF-1 and the abundant expression of growth factor receptors are not efforts for initiating an unrestrained cell proliferation, but rather they provide unliganded pathways for increased ER activation so as to strengthen estrogen signaling.

In nonobese patients, deficient estrogen signaling is associated with high risk for premature arteriosclerotic complications and cancer development. In these cases, inflammatory reactions at several sites represent compensatory actions against estrogen deficiency via an activation of aromatase enzyme. Moreover, in patients with clinically diagnosed infertility, PCOS, or BRCA mutation, the increased serum estrogen levels are compensatory efforts against ER resistance instead of a harmful induction of pathologic cell proliferation. Patients with ER resistance may exhibit an increased cancer risk in spite of hyperestrogenism, when the compensatory increased estrogen level is not sufficient for the breakthrough of ER blockade.

In conclusion, in obese patients, the causal therapy is the improvement of estrogen signaling in both men and women via estrogen substitution, changes in lifestyle, or use of natural products upregulating the genomic machinery. These therapies are equally protective against weight gain, metabolic disorders, inflammatory reactions, and obesity-related diseases including cancer. In obesity, medicaments working against the reparative, compensatory processes in the adipose tissue may induce strong counteractions via feedback mechanisms; however, the therapeutic results are transitory, ambiguous, and even dangerous.

Author details

Zsuzsanna Suba
Department of Molecular Pathology, National Institute of Oncology, Budapest, Hungary

*Address all correspondence to: subazdr@gmail.com

IntechOpen

References

[1] Nedungadi TP, Clegg DJ. Sexual dimorphism in body fat distribution and risk for cardiovascular diseases. Journal of Cardiovascular Translational Research. 2009;**2**(3):321-327

[2] Donohoe CL, Doyle SL, Reynolds JV. Visceral adiposity, insulin resistance and cancer risk. Diabetology and Metabolic Syndrome. 2011;**3**:12

[3] Palmer BF, Clegg DJ. The sexual dimorphism of obesity. Molecular and Cellular Endocrinology. 2015;**0**:113-119

[4] Davis SR, Castelo-Branco C, Chedraui P, Lumsden MA, Nappi RE, Shah D, et al. Understanding weight gain at menopause. Climacteric. 2012;**15**(5):419-429

[5] Lizcano F, Guzmán G. Estrogen deficiency and the origin of obesity during menopause. BioMed Research International. 2014;**2014**:757461

[6] Armani A, Berry A, Cirulli F, Caprio M. Molecular mechanisms underlying metabolic syndrome: The expanding role of the adipocyte. The FASEB Journal. 2017;**31**(10):4240-4255

[7] Bloomgarden ZT. Definitions of the insulin resistance syndrome: The 1st World Congress on the Insulin Resistance Syndrome. Diabetes Care. 2004;**27**(3):824-830

[8] Bloomgarden ZT. Second World Congress on the Insulin Resistance Syndrome: Mediators, pediatric insulin resistance, the polycystic ovary syndrome, and malignancy. Diabetes Care. 2005;**28**(8):1821-1830

[9] Suba Z. Common soil of smoking-associated and hormone-related cancers: Estrogen deficiency. Oncology Reviews. 2010;**4**(2):73-87

[10] Wierman ME. Sex steroid effects at target tissues: Mechanisms of action. Advances in Physiology Education. 2007;**31**:26-33

[11] Suba Z. DNA stabilization by the upregulation of estrogen signaling in BRCA gene mutation carriers. Drug Design, Development and Therapy. 2015;**9**:2663-2675

[12] Barakat R, Oakley O, Kim H, Jin J, Ko CJ. Extra-gonadal sites of estrogen biosynthesis and function. BMB Reports. 2016;**49**(9):488-496

[13] Arcidiacono B, Iiritano S, Nocera A, Possidente K, Nevolo MT, Ventura V. Insulin resistance and cancer risk: An overview of the pathogenetic mechanisms. Experimental Diabetes Research. 2012;**789174**:12

[14] Newell-Fugate AE. The role of sex steroids in white adipose tissue adipocyte function. Reproduction. 2017;**153**(4):R133-R149

[15] Suba Z. Circulatory estrogen level protects against breast cancer in obese women. Recent Patents on Anti-Cancer Drug Discovery. 2013;**8**(2):154-167

[16] Macotela Y, Boucher J, Tran TT, Kahn CR. Sex and depot differences in adipocyte insulin sensitivity and glucose metabolism. Diabetes. 2009;**58**:803-812

[17] Nestler JE. Obesity, insulin, sex steroids and ovulation. International Journal of Obesity and Related Metabolic Disorders. 2000;**24** (Suppl 2):S71-S73

[18] Suba Z. Diverse pathomechanisms leading to the breakdown of cellular estrogen surveillance and breast cancer development: New therapeutic strategies. Drug Design, Development and Therapy. 2014;**8**:1381-1390

[19] Steinbaum SR. The metabolic syndrome: An emerging health epidemic in women. Progress in Cardiovascular Diseases. 2004;**46**(4):321-336

[20] Ryan AS, Nicklas BJ, Berman DM. Hormone replacement therapy, insulin sensitivity, and abdominal obesity in postmenopausal women. Diabetes Care. 2002;**25**:127-133

[21] Smith EP, Boyd J, Frank GR, Takahashi H, Cohen RM, Specker B, et al. Estrogen resistance caused by a mutation in the estrogen-receptor gene in a man. The New England Journal of Medicine. 1994;**331**:1056-1061

[22] Quaynor SD, Stradtman EW, Kim HG, Shen Y, Chorich LP, Schreihofer DA, et al. Delayed puberty and estrogen resistance in a woman with estrogen receptor α variant. The New England Journal of Medicine. 2013;**369**(2):164-171

[23] Morishima A, Grumbach MM, Simpson ER, Fisher C, Qin K. Aromatase deficiency in male and female siblings caused by a novel mutation and the physiological role of estrogens. The Journal of Clinical Endocrinology and Metabolism. 1995;**80**(12):3689-3698

[24] Choi SB, Jang JS, Park S. Estrogen and exercise may enhance beta-cell function and mass via insulin receptor substrate-2 induction in ovariectomized diabetic rats. Endocrinology. 2005;**146**(11):4786-4794

[25] Wade GN, Gray JM. Gonadal effects on food intake and adiposity: A metabolic hypothesis. Physiology & Behavior. 1979;**22**:583-593

[26] Jones ME, Thorburn AW, Britt KL, et al. Aromatase-deficient (ArKO) mice have a phenotype of increased adiposity. Proceedings of the National Academy of Sciences of the United States of America. 2000;**97**:12735-12740

[27] Heine PA, Taylor JA, Iwamoto GA, Lubahn DB, Cooke PS. Increased adipose tissue in male and female estrogen receptor-α knockout mice. Proceedings of the National Academy of Sciences of the United States of America. 2000;**97**:12729-12734

[28] Nkhata KJ, Ray A, Dogan S, Grande JP, Cleary MP. Mammary tumor development from T47-D human breast cancer cells in obese ovariectomized mice with and without estradiol supplements. Breast Cancer Research and Treatment. 2009;**114**(1):71-83

[29] Hong J, Holcomb VB, Kushiro K, Núñez NP. Estrogen inhibits the effects of obesity and alcohol on mammary tumors and fatty liver. International Journal of Oncology. 2011;**39**(6):1443-1453

[30] Renehan AG, Tyson M, Egger M, et al. Body-mass index and incidence of cancer: A systematic review and meta-analysis of prospective observational studies. Lancet. 2008;**371**(9612):569-578

[31] Suba Z. Hormonal and metabolic disorders as links between obesity and cancer risk in men and women. International Journal of Cancer Research and Prevention. 2016;**9**(4):335-352

[32] Mathieu LN, Kanarek NF, Tsai HL, Rudin CM, Brock MV. Age and sex differences in the incidence of esophageal adenocarcinoma: Results from the Surveillance, Epidemiology, and End Results (SEER) Registry (1973-2008). Diseases of the Esophagus. 2014;**27**(8):757-763

[33] Suba Z. Gender-related hormonal risk factors for oral cancer. Pathology Oncology Research. 2007;**13**:195-202

[34] Yang P, Zhou Y, Chen B, Wan HW, Jia GQ, Bai HL, et al. Overweight, obesity and gastric cancer risk: Results from a meta-analysis of cohort

studies. European Journal of Cancer. 2009;**45**(16):2867-2873

[35] Kennelly R, Kavanagh DO, Hogan AM, Winter DC. Oestrogen and the colon: Potential mechanisms for cancer prevention. The Lancet Oncology. 2008;**9**(4):385-391

[36] Aleksandrova K, Stelmach-Mardas M, Schlesinger S. Obesity and Liver Cancer. Recent Results in Cancer Research. 2016;**208**:177-198

[37] Chow WH, McLaughlin JK, Mandel JS, et al. Obesity and risk of renal cell cancer. Cancer Epidemiology, Biomarkers & Prevention. 1996;**5**(1):17-21

[38] Woldrich JM, Mallin K, Ritchey J, Carroll PR, Kane CJ. Sex differences in renal cell cancer presentation and survival: An analysis of the National Cancer Database, 1993-2004. The Journal of Urology. 2008;**179**(5):1709-1713

[39] Weiss JR, Moysich KB, Swede H. Epidemiology of male breast cancer. Cancer Epidemiology, Biomarkers & Prevention. 2005;**14**(1):20-26

[40] Hsing AW, McLaughlin JK, Cocco P, Co Chien HT, Fraumeni JF Jr. Risk factors for male breast cancer (United States). Cancer Causes & Control. 1998;**9**(3):269-275

[41] Rose DP, Vona-Davis L. Interaction between menopausal status and obesity in affecting breast cancer risk. Maturitas. 2010;**66**(1):33-38

[42] Suba Z. Triple-negative breast cancer risk in women is defined by the defect of estrogen signaling: Preventive and therapeutic implications. OncoTargets and Therapy. 2014;**7**:147-164

[43] Lahmann PH, Hoffmann K, Allen N, van Gils CH, Khaw KT, Tehard B, et al. Body size and breast cancer risk: Findings from the European Prospective Investigation into Cancer and Nutrition (EPIC). International Journal of Cancer. 2004;**111**(5):762-771

[44] Huang Z, Willett WC, Colditz GA, Hunter DJ, Manson JE, Rosner B, et al. Waist circumference, waist:hip ratio, and risk of breast cancer in the Nurses' Health Study. American Journal of Epidemiology. 1999;**150**(12):1316-1324

[45] Morimoto LM, White E, Chen Z, Chlebowski RT, Hays J, Kuller L, et al. Obesity, body size and risk of postmenopausal breast cancer: The Women's Health Initiative (United States). Cancer Causes & Control. 2002;**13**(8):41-51

[46] Kershaw EE, Flier JS. Adipose tissue as an endocrine organ. The Journal of Clinical Endocrinology and Metabolism. 2004;**89**(6):2548-2556

[47] Wang P, Mariman E, Renes J, Keijer J. The secretory function of adipocytes in the physiology of white adipose tissue. Journal of Cellular Physiology. 2008;**216**:3-13

[48] Maggi A. Liganded and unliganded activation of estrogen receptor and hormone replacement therapies. Biochimica et Biophysica Acta. 2011;**1812**(8):1054-1060

[49] Dieudonné MN, Leneveu MC, Giudicelli Y, Pecquery R. Evidence for functional estrogen receptors α and β in human adipose cells: Regional specificities and regulation by estrogens. American Journal of Physiology-Cell Physiology. 2004;**286**(3):C655-C661

[50] Tao Z, Zheng LD, Smith C, Luo J, Robinson A, Almeida FA, et al. Estradiol signaling mediates gender difference in

visceral adiposity via autophagy. Cell Death & Disease. 2018;**9**(3):309

[51] Hemsell DL, Grodin JM, Brenner PF, Siiteri PK, MacDonald PC. Plasma precursors of estrogen. II. Correlation of the extent of conversion of plasma androstenedione to estrone with age. The Journal of Clinical Endocrinology and Metabolism. 1974;**38**:476-479

[52] Björnström L, Sjöberg M. Mechanisms of estrogen receptor signaling: Convergence of genomic and nongenomic actions on target genes. Molecular Endocrinology. 2005;**19**(4):833-842

[53] Nelson LR, Bulun SE. Estrogen production and action. Journal of the American Academy of Dermatology. 2001;**45**:S116-S124

[54] Labrie F, Bélanger A, Luu-The V, et al. DHEA and the intracrine formation of androgens and estrogens in peripheral target tissues: Its role during aging. Steroids. 1998;**63**:322-328

[55] Suba Z. Activating mutations of ESR1, BRCA1 and CYP19 aromatase genes confer tumor response in breast cancers treated with antiestrogens. Recent Patents on Anti-Cancer Drug Discovery. 2017;**12**(2):136-147

[56] Blüher M. Importance of estrogen receptors in adipose tissue function. Molecular Metabolism. 2013;**2**:130-132

[57] Saltiel AR, Kahn CR. Insulin signaling and the regulation of glucose and lipid metabolism. Nature. 2001;**414**:799-806

[58] Suba Z. Low estrogen exposure and/or defective estrogen signaling induces disturbances in glucose uptake and energy expenditure. Journal of Diabetes and Metabolism. 2013;**4**:272-281

[59] Muraki K, Okuya S, Tanizawa Y. Estrogen receptor α regulates insulin sensitivity through IRS-1 tyrosine phosphorylation in mature 3T3-L1 adipocytes. Endocrine Journal. 2006;**53**(6):841-851

[60] Kim J, Jo K, Kim B, Baik H, Lee S. 17 beta-estradiol induces an interaction between adenosine monophosphate-activated protein kinase and the insulin signaling pathway in 3T3-L1 adipocytes. International Journal of Molecular Medicine. 2012;**30**:979-985

[61] Campello RS, Fátima LA, Barreto-Andrade JN, Lucas TF, Mori RC, Porto CS, et al. Estradiol-induced regulation of GLUT4 in 3T3-L1 cells: Involvement of ESR1 and AKT activation. Journal of Molecular Endocrinology. 2017;**59**(3):257-268

[62] Rosenbaum D, Haber RS, Dunaif A. Insulin resistance in polycystic ovary syndrome: Decreased expression of GLUT-4 glucose transporters in adipocytes. American Journal of Physiology. 1993;**264**(2 Pt 1):E197-E202

[63] Gorres BK, Bomhoff GL, Gupte AA, Geiger PC. Altered estrogen receptor expression in skeletal muscle and adipose tissue of female rats fed a high-fat diet. Journal of Applied Physiology. 2011;**110**:1046-1053

[64] Nilsson M, Dahlman I, Rydén M, et al. Oestrogen receptor α gene expression levels are reduced in obese compared to normal weight females. International Journal of Obesity. 2007;**31**:900-907

[65] Zhang H, Chen X, Sairam MR. Novel genes of visceral adiposity: Identification of mouse and human mesenteric estrogen-dependent adipose (MEDA)-4 gene and its adipogenic function. Endocrinology. 2012;**153**:2665-2676

[66] Suba Z. Interplay between insulin resistance and estrogen deficiency as co-activators in carcinogenesis. Pathology Oncology Research. 2012;**18**(2):123-133

[67] Barros RP, Gustafsson JÅ. Estrogen receptors and the metabolic network. Cell Metabolism. 2011;**14**:289-299

[68] Parthasarathy C, Renuka VN, Balasubramanian K. Sex steroids enhance insulin receptors and glucose oxidation in Chang liver cells. Clinica Chimica Acta. 2009;**399**:49-53

[69] Iavarone M, Lampertico P, Seletti C, et al. The clinical and pathogenetic significance of estrogen receptor-beta expression in chronic liver diseases and liver carcinoma. Cancer. 2003;**98**:529-534

[70] Klair JS, Yang JD, Abdelmalek MF, et al. A longer duration of estrogen deficiency increases fibrosis risk among postmenopausal women with nonalcoholic fatty liver disease. Hepatology. 2016;**64**:85-91

[71] Yonezawa R, Wada T, Matsumoto N, et al. Central versus peripheral impact of estradiol on the impaired glucose metabolism in ovariectomized mice on a high-fat diet. American Journal of Physiology. 2012;**303**:E445-E456

[72] Stubbins RE, Najjar K, Holcomb VB, Hong J, Núñez NP. Oestrogen alters adipocyte biology and protects female mice from adipocyte inflammation and insulin resistance. Diabetes, Obesity and Metabolism. 2012;**14**(1):58-66

[73] Sukocheva OA. Estrogen, estrogen receptors, and hepatocellular carcinoma: Are we there yet? World Journal of Gastroenterology. 2018;**24**(1):1-4

[74] Marzioni M, Torrice A, Saccomanno S, et al. An oestrogen receptor beta-selective agonist exerts antineoplastic effects in experimental intrahepatic cholangiocarcinoma. Digestive and Liver Disease. 2012;**44**:134-142

[75] McGlynn KA, Sahasrabuddhe VV, Zeleniuch-Jacquotte A. Reproductive factors, exogenous hormone use and risk of hepatocellular carcinoma among US women: Results from the Liver Cancer Pooling Project. British Journal of Cancer. 2015;**112**:1266-1272

[76] Yuchi Y, Cai Y, Legein B, et al. Estrogen receptor alpha regulates beta-cell formation during pancreas development and following injury. Diabetes. 2015;**64**:3218-3228

[77] Le May C, Chu K, Hu M, et al. Estrogens protect pancreatic beta-cells from apoptosis and prevent insulin-deficient diabetes mellitus in mice. Proceedings of the National Academy of Sciences of the United States of America. 2006;**103**:9232-9237

[78] Alonso-Magdalena P, Ropero AB, Carrera MP, et al. Pancreatic insulin content regulation by the estrogen receptor ER alpha. PLoS One. 2008;**3**:e2069

[79] Tiano JP, Delghingaro-Augusto V, Le May C, Liu S, Kaw MK, Khuder SS, et al. Estrogen receptor activation reduces lipid synthesis in pancreatic islets and prevents β cell failure in rodent models of type 2 diabetes. The Journal of Clinical Investigation. 2011;**121**(8):3331-3342

[80] Guan HB, Wu L, Wu QJ, Zhu J, Gong T. Parity and pancreatic cancer risk: A dose-response meta-analysis of epidemiologic studies. PLoS One. 2014;**9**(3):e92738

[81] Wada-Hiraike O, Imamov O, Hiraike H, et al. Role of estrogen receptor beta in colonic epithelium. Proceedings of the National Academy of Sciences of the United States of America. 2006;**103**:2959-2964

[82] Choijookhuu N, Hino S-I, Oo PS, Batmunkh B, Hishikawa Y. The role of estrogen receptors in intestinal homeostasis and disease. Receptors & Clinical Investigation. 2016;3:e1109

[83] Barlow GM, Yu A, Mathur R. Role of the gut microbiome in obesity and diabetes mellitus. Nutrition in Clinical Practice. 2015;30(6):787-797

[84] Baker JM, Al-Nakkash L, Herbst-Kralovetz MM. Estrogen-gut microbiome axis: Physiological and clinical implications. Maturitas. 2017;103:45-53

[85] Marotz CA, Zarrinpar A. Treating obesity and metabolic syndrome with fecal microbiota transplantation. The Yale Journal of Biology and Medicine. 2016;89(3):383-388

[86] Gluhovschi G, Gluhovschi A, Anastasiu D, Petrica L, Gluhovschi C, Velciov S. Chronic kidney disease and the involvement of estrogen hormones in its pathogenesis and progression. Romanian Journal of Internal Medicine. 2012;50(2):135-144

[87] Liu H, Wang XC, Hu GH, Huang TB, Xu YF. Oral contraceptive use and kidney cancer risk among women: Evidence from a meta-analysis. International Journal of Clinical and Experimental Medicine. 2014;7(11):3954-3963

[88] Reckelhoff JF. Sex steroids, cardiovascular disease, and hypertension: Unanswered questions and some speculations. Hypertension. 2005;45(2):170-174

[89] Nestler JE, Jakubowicz DJ, deVargas AF, Brik C, Quintero N, Medina F. Insulin stimulates testosterone biosynthesis by human thecal cells from women polycystic ovary syndrome by activating its own receptor and using inositolglycan mediators as the signal transduction system. The Journal of Clinical Endocrinology and Metabolism. 1998;83(6):2001-2005

[90] Moghetti P, Castello R, Negri C, Tosi F, Spiazzi GG, Brun E, et al. Insulin infusion amplifies 17 alpha-hydroxycorticosteroid intermediates response to adrenocorticotropin in hyperandrogenic women: Apparent relative impairment of 17,20-lyase activity. The Journal of Clinical Endocrinology and Metabolism. 1996;81(3):881-886

[91] Cara JF, Fan J, Azzarello J, Rosenfield RL. Insulin-like growth factor-I enhances luteinizing hormone binding to rat ovarian theca-interstitial cells. The Journal of Clinical Investigation. 1990;86(2):560-565

[92] Stamataki KE, Spina J, Rangou DB, Chlouverakis CS, Piaditis GP. Ovarian function in women with non-insulin dependent diabetes mellitus. Clinical Endocrinology. 1996;45(5):615-621

[93] Walsh BW, Schiff I, Rosner B, Greenberg L, Ravnikar V, Sacks FM. Effects of postmenopausal estrogen replacement on the concentrations and metabolism of plasma lipoproteins. The New England Journal of Medicine. 1991;325(17):1196-1204

[94] Harrison-Bernard LM, Schulman IH, Raij L. Postovariectomy hypertension is linked to increased renal AT1 receptor and salt sensitivity. Hypertension. 2003;42(6):1157-1163

[95] Widder J, Pelzer T, von Poser-Klein C, Hu K, Jazbutyte V, Fritzemeier KH, et al. Improvement of endothelial dysfunction by selective estrogen receptor-alpha stimulation in ovariectomized SHR. Hypertension. 2003;42(5):991-996

[96] Pare G, Krust A, Karas RH, Dupont S, Aronovitz M, Chambon P, et al. Estrogen receptor-alpha mediates the protective effects of estrogen against

vascular injury. Circulation Research. 2002;**90**(10):1087-1092

[97] Giordano SH, Kuo Y-F, Freeman JL, Buchholz TA, Hortobagyi GN, Goodwin JS. Risk of cardiac death after adjuvant radiotherapy for breast cancer. Journal of the National Cancer Institute. 2005;**97**(6):419-424

[98] Suba Z. Causal therapy of breast cancer irrelevant of age, tumor stage and er-status: Stimulation of estrogen signaling coupled with breast conserving surgery. Recent Patents on Anti-Cancer Drug Discovery. 2016;**11**(3):254-266

[99] Clegg DJ, Brown LM, Woods SC, Benoit SC. Gonadal hormones determine sensitivity to central leptin and insulin. Diabetes. 2006;**55**:978-987

[100] Simpson ER, Misso M, Hewitt KN, et al. Estrogen—The good, the bad, and the unexpected. Endocrine Reviews. 2005;**26**(3):322-330

[101] Combs TP, Pajvani UB, Berg AH, et al. A transgenic mouse with a deletion in the collagenous domain of adiponectin displays elevated circulating adiponectin and improved insulin sensitivity. Endocrinology. 2004;**145**(1):367-383

[102] Steppan CM, Bailey ST, Bhat S, et al. The hormone resistin links obesity to diabetes. Nature. 2001;**409**(6818):307-312

[103] De Pergola G, Silvestris F. Obesity as a major risk factor for cancer. Journal of Obesity. 2013;**2013**:291546

[104] Huh JY, Park YJ, Ham M, Kim JB. Crosstalk between adipocytes and immune cells in adipose tissue inflammation and metabolic dysregulation in obesity. Molecules and Cells. 2014;**37**(5):365-371

[105] Trayhurn P, Wood IS. Adipokines: Inflammation and the pleiotropic role of

white adipose tissue. The British Journal of Nutrition. 2004;**92**(3):347-355

[106] Purohit A, Newman SP, Reed MJ. The role of cytokines in regulating estrogen synthesis: Implications for the etiology of breast cancer. Breast Cancer Research. 2002;**4**:65

[107] Ye J, McGuinness OP. Inflammation during obesity is not all bad: Evidence from animal and human studies. American Journal of Physiology-Endocrinology and Metabolism. 2013;**304**(5):E466-E477

[108] Boucher J, Tseng J-H, Kahn CR. Insulin and insulin-like growth factor receptors act as ligand-specific amplitude modulators of a common pathway regulating gene. The Journal of Biological Chemistry. 2010;**285**(22):17235-17245

[109] Lukanova A, Söderberg S, Stattin P, Palmqvist R, Lundin E, Biessy C, et al. Nonlinear relationship of insulin-like growth factor (IGF)-I and IGF-I/IGF-binding protein-3 ratio with indices of adiposity and plasma insulin concentrations (Sweden). Cancer Causes & Control. 2002;**13**(6):509-516

[110] Smith CL. Cross-talk between peptide growth factor and estrogen receptor signaling pathways. Biology of Reproduction. 1998;**58**(3):627-632

[111] Xu J, Xiang Q, Lin G, Fu X, Zhou K, Jiang P, et al. Estrogen improved metabolic syndrome through down-regulation of VEGF and HIF-1α to inhibit hypoxia of periaortic and intra-abdominal fat in ovariectomized female rats. Molecular Biology Reports. 2012;**39**(8):8177-8185

[112] Caizzi L, Ferrero G, Cutrupi S, Cordero F, Ballaré C, Miano V, et al. Genome-wide activity of unliganded estrogen receptor-α in breast cancer cells. Proceedings of the National Academy of Sciences

of the United States of America.
2014;**111**(13):4892-4897

[113] Jordan VC. The new biology
of estrogen-induced apoptosis
applied to treat and prevent breast
cancer. Endocrine-Related Cancer.
2015;**22**(1):R1-R31

[114] Suba Z. The pitfall of the transient,
inconsistent anticancer capacity of
antiestrogens and the mechanism of
apparent antiestrogen resistance. Drug
Design, Development and Therapy.
2015;**9**:4341-4353

[115] Suba Z. Amplified crosstalk
between estrogen binding and GFR
signaling mediated pathways of ER
activation drives responses in tumors
treated with endocrine disruptors.
Recent Patents on Anti-Cancer Drug
Discovery. 2018;**13**(4):428-444

Chapter 3

Oral Health and Cardiovascular Disorders

Ioana Mozos and Dana Stoian

Abstract

Several studies reported the cross talk between oral health and cardiovascular disorders. The aim of the present chapter is to review the main mechanisms linking oral and cardiovascular disorders, the main pathologies which could be linked, and possibilities for prophylactic and therapeutic interventions. Periodontitis was associated with cardiovascular risk, and the links between the two entities are represented by bacteria and their toxins released into the blood, causing endothelial dysfunction and providing a proatherogenic and prothrombotic effect and an inflammatory and immune reaction. The mentioned mechanisms explain the reported associations of periodontitis with stroke, coronary heart disease, and peripheral vascular disease. Periodontitis was also associated with diabetes mellitus and impaired lipid metabolism. Not all studies confirmed the association between periodontitis and coronary artery disease or stroke. Tooth loss, the most important consequence of periodontitis, has been also associated with cardiovascular disease. Dental and pulpal caries were also found to be independent risk factors for atherosclerosis, while restorations were inversely related to an atherosclerotic burden. Sucrose is involved in both cariogenesis and atherosclerosis. Fluorides prevent aortic calcifications and enamel demineralization and inhibit bacterial metabolism but are cardiotoxic. Heightening awareness of good dental hygiene can improve cardiovascular health.

Keywords: periodontitis, periodontal bacteria, dental caries, tooth loss, sucrose, fluorides, cardiovascular risk, inflammation, dyslipidemia, abdominal aorta aneurysm

1. Introduction

Cardiovascular disorders are the main mortality causes worldwide. The most common substrate is atherosclerosis, with several risk factors, a latent evolution, and possible sudden, unexpected fatal outcomes. Oral diseases, the most common chronic diseases, are important public health problems considering their prevalence, their impact on health and quality of life, and therapy expenses [1].

Several links have been reported between cardiovascular and oral disorders. The present chapter aims to review and emphasize the main mechanisms linking oral and cardiovascular disorders, the main pathologies which could be linked, and possibilities for prophylactic and therapeutic interventions.

2. Periodontitis (PD)

Periodontitis, the chronic inflammatory disease irreversibly affecting the supporting structures of the teeth, occurs due to periodontopathogen bacteria and subgingival plaque [2, 3]. Periodontitis was associated with an impaired cardiovascular health, endothelial dysfunction, and atherosclerosis [2, 4]. Several epidemiological studies, review articles, and meta-analysis confirmed a positive association between periodontitis and cardiovascular disorders [5–8].

2.1 Periodontal bacteria

Different species of periodontal bacteria cause periodontitis, including *Porphyromonas gingivalis*, *Aggregatibacter actinomycetemcomitans*, and *Prevotella intermedia* [9]. A synergistic, physiologically compatible, and dysbiotic microbial community, including keystone pathogens such as *Porphyromonas gingivalis*, impairs and manipulates the defense mechanisms of the host (delaying recruitment or destroying neutrophils, degradation of antibodies and complement) and increases the virulence of the entire microbial community [10]. Bacteremia is caused by chewing, tooth brushing, dental flossing, or invasive dental maneuvers, such as tooth extraction or periodontal treatment [11, 12]. The intensity of bacteremia depends on the severity of periodontitis [8]. The oral cavity and periodontal pockets represent reservoirs of gram-negative and anaerobic bacteria, able to interact with the cardiovascular tissue [11]. *Porphyromonas gingivalis* and *Prevotella intermedia* generate antibodies, which can also cause tissue injury [13]. Antibodies against *Porphyromonas gingivalis* cross-react with heat shock proteins expressed by endothelial cells, contributing to detrimental cardiovascular consequences [14]. Molecular mimicry between human and bacterial heat shock proteins enables also atherogenesis [3]. Periodontal pathogens and their noxious products are released into the systemic circulation through the ulcerated sulcular epithelium of the gingiva, and they can induce insulin resistance, systemic inflammation, are involved in all stages of atherogenesis, and explain, probably, the perio-systemic link through the released inflammatory mediators [15, 16]. Dissemination through the blood of alive periodontal pathogens is possible also through immune cells [3].

Porphyromonas gingivalis can invade aortic endothelial cells, due to their fimbriae, and fimbrillin peptides can induce the expression of interleukin 8 and monocyte chemotactic protein [17]. *Porphyromonas gingivalis* multiply inside the endothelial cells and activate Toll-like receptor 2, resulting in release of pro-inflammatory cytokines [8]. *Porphyromonas gingivalis*-infected endothelial cells have also a higher expression of adhesion molecules, which exert a proatherogenic effect [17]. Bacteria and their virulence factors directly stimulate white blood cells, fibroblasts, mast, and dendritic and endothelial cells, causing an inflammatory reaction and expression of metalloproteinases [14, 18]. Oral bacteria such as *Porphyromonas gingivalis*, *Aggregatibacter actinomycetemcomitans*, and *Prevotella intermedia* and oral bacteria DNA and RNA have also been identified in atheroma, and they are able to activate monocytes, which transform into macrophages and foam cells [8, 14, 18, 19]. Plaque rupture is caused by endothelial cell apoptosis and extracellular matrix degradation due to periodontal pathogens [8].

Toxins of periodontal pathogens, such as proteases (gingipain), adhesins, and lectins, regulate the bacterial biofilm; can also impair the immune response of the host, by degradation of interleukins; and enable atherosclerotic plaque formation by increasing proliferation of vascular smooth muscle cells and platelet aggregation [3].

Probiotics might represent a solution, preventing dental caries and reducing episodes of streptococcal pharyngeal infection, pocket depth, and plaque index [14]. Several cardiovascular benefits have been also described for probiotics, such as lowering blood pressure and circulating lipids [20].

2.2 Inflammatory reaction linking periodontitis and cardiovascular disorders

Periodontitis is a source of inflammatory mediators, such as TNF alpha and interleukin 1 and 6 [2]. Inflammatory markers may be produced locally in the oral cavity and released into the bloodstream or are the result of bacteremia [14].

Atherosclerosis is a high inflammatory state, and inflammatory mediators can cause endothelial dysfunction, a proatherogenic and prothrombotic effect, and are involved in the development and evolution of the atherosclerotic plaque [18, 21]. The same inflammatory markers are involved in both atherosclerosis and periodontitis [21]. Periodontitis impairs endothelial function and is also associated with oxidative stress [13].

2.3 Periodontitis, cardiovascular risk, and disorders

Periodontitis was associated with several cardiovascular risk factors, especially hypertension, diabetes mellitus, dyslipidemia, and smoking [21]. Smoking is a risk factor for both cardiovascular disorders and oral health. It is known to impair periodontal tissue perfusion, inflammatory reaction, fibroblast function, and tissue healing, causing also endothelial dysfunction [2]. Several cardiovascular disorders were associated with periodontitis, such as stroke, coronary heart disease, peripheral vascular disease, arrhythmias, and aorta aneurysm [11, 13, 22].

A link between periodontitis and abdominal aorta aneurysm has been demonstrated [11, 23]. Periodontal pathogens or their by-products may be involved in systemic and local mechanisms associated with the development and evolution of abdominal aorta aneurysm (AAA), suggesting an infective theory of AAA pathogenesis. [23–26]. There is, probably, a translocation of periodontal pathogens from subgingival microbiota to the bloodstream and then to the aortic wall, where they might contribute to the weakening of the aortic wall or secondary colonize AAAs. [9, 23, 27]. Suzuki et al. reported deeper pocket depth and more severe bleeding on probing in abdominal aorta aneurysm patients compared to the non-AAA patients [9] (**Table 1**).

Periodontopathic bacteria were detected in aortic walls of patients with AAA [28]. *Porphyromonas gingivalis* worsened AAA development through toll-like receptor signaling and matrix metalloproteinases (MMPs) [28]. Pharmacological inactivation of MMPs prevented the development of AAA [29]. Several other periodontal pathogens were also involved in the pathogenesis of AAA, such as *Treponema denticola* and *Campylobacter rectus* [11].

The impaired blood flow through vessels affected by atherosclerosis may affect also the gums and explain the association of coronary heart disease or peripheral arterial disease with periodontitis [30]. A poor periodontal state with increased inflammatory markers (C-reactive protein and TNF alpha) was reported in a study including 25 patients with peripheral arterial disease [22]. Kure et al. described a mutual inflammatory pathway in patients with both periodontitis and peripheral arterial disease, including local and systemic mechanisms [22]. The association of oral health-myocardial infarction was explained by common factors, such as low socioeconomic status, smoking,

Study participants	Assessed variables	Results	Conclusions	Reference
142 patients with tachycardia (TA) and 25 patients with AAA	• Periodontitis • *P. gingivalis*, *A. actinomy*, *Prevotella intermedia* in saliva and subgingival plaque (PCR) • Serum antibody titers against the bacteria (ELISA)	• Fewer remaining teeth and deeper pocket depth in AAA patients than in TA patients; comparable periodontal bacteria and antibody titers between the two groups	• Periodontitis has a more important influence on aneurysm progression than on arrhythmia	Suzuki et al. [25]
12 patients with AAA and 24 sex- and age-matched non-AAA patients	• Periodontal pathogens: *P. gingivalis*, *A. actinomy*, and *Prevotella intermedia* in oral samples using PCR • Probing pocket depth (PD), bleeding on probing (BOP), number of teeth, community periodontal index (CPI)	• AAA patients had deeper PD, more severe BOP, average CPI, and less number of teeth; periodontal bacteria were not different between the two groups	• AAA patients have bad oral and periodontal conditions	Suzuki et al. [9]
70 patients with PAD or AAA	• Periodontal pathogens (*A. actinomy*, *P. gingivalis*, *Campylobacter rectus*, and *Tannerella forsythia*) in vascular, blood, and subgingival samples (quantitative and nested PCR)	• All periodontal pathogens were found in subgingival sample • Bacterial DNA was detected in a few patients in vascular and blood samples	• There is probably a migration of periodontal pathogens from subgingival microbiota to the bloodstream and to atheromatous plaques	Figuero et al. [27]
32 AA patients	• 7 periodontal bacteria (PCR) in aneurysmal walls, mural thrombi, occlusive atherosclerotic aorta, and control arterial tissue	• Bacteria were found in the aneurysmal walls and occlusive aorta	• Bacteria may play a role in development or weakening of the aneurysmal wall in AAA	Kurihara et al. [11]

Note: P. gingivalis = Porphyromonas gingivalis, A. actinomy = Aggregatibacter actinomycetemcomitans, PCR = polymerase chain reaction assays, PAD = peripheral arterial disease.

Table 1.
Periodontitis and abdominal aorta aneurysm (AAA).

diabetes mellitus, insulin resistance, nutrition, and aging. Persons who care for dental health pay also special attention to lifestyle.

Periodontitis was also associated with myocardial hypertrophy [12]. Ventricular remodeling after myocardial infarction and pressure overload myocardial hypertrophy were aggravated by *Aggregatibacter actinomycetemcomitans* [12, 31]. Periodontal pathogens increase the level of MMPs, especially MMP-2, which are responsible for gingival extracellular matrix destruction [12]. MMPs are also released into the blood and are responsible for occurrence of myocardial inflammation, pressure overload myocardial hypertrophy, myocardial interstitial, and perivascular fibrosis, causing systolic and diastolic dysfunction [12].

Significant associations were also reported between periodontitis and cardiac arrhythmia. Periodontitis, as a chronic inflammatory disorder, was an independent predictor of arrhythmic events and major adverse cardiac events in patients with atrial fibrillation (AF) according to a study including 227 patients with AF [13]. The link periodontitis-atrial fibrillation is explained by the inflammatory infiltration of atrial cardiomyocytes causing hypertrophy, oxidative stress, and the cardiac tissue damage induced by antibodies generated in response to *Porphyromonas gingivalis* and *Prevotella intermedia* [13, 32]. The control of periodontitis may improve inflammation and may prevent embolic events and recurrence of arrhythmia [13]. Chen et al. also reported an increased risk of atrial fibrillation or flutter in patients with periodontitis, according to a study including the nationwide database of the Taiwanese population (393,745 patients with periodontitis and 393,745 non-PD individuals) [33]. However, the association PD-atrial fibrillation or flutter was not significant in patients with hyperthyroidism, a disorder with a significant role in triggering those arrhythmias [33]. Other arrhythmias, such as atrial tachycardia, atrial and ventricular premature contractions, and ventricular tachycardia, were also more common in patients with periodontitis [13]. Periodontitis may influence tachyarrhythmia progression [34]. *Porphyromonas gingivalis* and *Prevotella intermedia* were especially detected in saliva from tachyarrhythmia patients (compared to bradyarrhythmia subjects), and they may be involved in myocardial remodeling [34].

2.4 Periodontitis and dyslipidemia

Several researchers considered the potential relationship between periodontitis and lipid metabolism [35]. According to a meta-analysis including 19 studies, with nonsmoking subjects, with chronic periodontitis, not taking any lipid lowering drugs, periodontitis was significantly associated with low HDL and high LDL and triglyceride levels [35]. The link between periodontitis and low HDL could be periodontal infection and inflammation [36]. Certain bacteria can also increase VLDL and small dense LDL levels and induce selective proteolysis of apoprotein B-100, supporting the role of lipoproteins in linking periodontitis and atherosclerosis [3].

Not all studies confirmed the link between lipid metabolism disorders and periodontal health. Sridhar et al. reported no association between periodontal health, assessed using the gingival index, oral hygiene index, periodontal disease index scores and attachment loss, and serum lipids (total cholesterol, LDL, HDL, and triglycerides) in patients with or without coronary heart disease [37] (**Table 1**). Another study, including 1297 nondiabetic subjects, who had never smoked and were under 50 years, a subpopulation of the Health 2000 Health Examination Survey, revealed a significant association of high serum triglycerides and low HDL with periodontal infection only among obese patients [38]. TNF alpha may increase triglycerides by impairing lipoprotein lipase activity [18].

Proprotein convertase subtilisin/kexin type 9 (PCSK9) plays a critical role in regulation of circulating LDL cholesterol concentrations and seems to be upregulated in patients with periodontitis, according to a study including 40 periodontitis patients with *Porphyromonas gingivalis* antibodies [39] (**Table 2**). In Japanese male subjects, the concentrations of serum proprotein convertase subtilisin/kexin type 9 (PCSK9) correlated with periodontal parameters according to a study including 108 male subjects [40] (**Table 2**).

A bidirectional relationship periodontitis-dyslipidemia was suggested, considering that periodontal inflammation impairs lipid metabolism and dyslipidemia increases the susceptibility for periodontitis due to the associated inflammatory reaction [3].

2.5 Periodontitis and hypertension

Periodontitis was associated with poor blood pressure control, especially in elderly patients; a good periodontal health enabled improvement of systolic blood pressure through antihypertensive therapy [41, 42]. Severity of periodontitis has also been associated with increased systolic blood pressure [43]. The number of periodontal pockets was significantly related to hypertension in a study including

Study participants	Results	Conclusions	Reference
108 male subjects with and without periodontal disease	The concentrations of PCSK9 (proprotein convertase subtilisin/kexin type 9) were increased in subjects with periodontal disease (determined as a probing depth of $\geq$4 mm in at least one site) compared with healthy subjects	PCSK9 could be a biomarker for diagnosing chronic periodontitis and may evaluate the risk for periodontitis to cause atherosclerotic vascular disease	Tabeta et al. [40]
40 periodontitis patients and 30 control subjects	PCSK9 concentrations in periodontitis patients were significantly higher than in control subjects. No correlations were found between PCSK9 concentrations and lipid profiles	Periodontal infection upregulates PCSK9 production	Miyazawa et al. [39]
120 subjects, with 30 subjects in each group: healthy group (A), chronic periodontitis group (B), coronary heart disease (CHD + periodontitis group) (C), and CHD – periodontitis group (D)	No significant difference with respect to the lipid profile levels (total cholesterol, HDL, LDL, triglycerides) was noticed between the four groups	Periodontal disease did not cause any change in the lipid profile in systemically healthy or in CHD patients	Sridhar et al. [37]
1297 dentate, nondiabetic subjects, nonsmokers, aged under 50 years	No consistent association was found between serum lipid levels (HDL, LDL, and triglycerides) and periodontal infection among normoweight subjects but an association of high serum triglycerides and low HDL with periodontal infection among obese subjects	Lipid levels were associated with periodontal infection only in obese subjects with a high serum triglyceride level and/or a low HDL-cholesterol level	Saxlin et al. [38]

Table 2.
Studies linking lipid metabolism and periodontitis.

3352 patients with self-reported history of hypertension and myocardial infarction [44]. Intensive periodontal treatment, including the use of a locally delivered antimicrobial, reduced systemic inflammatory markers and systolic blood pressure and improved lipid profiles and cardiovascular risk in 40 patients with severe periodontitis, in a 6-month trial [43].

2.6 Periodontitis and diabetes mellitus

There is a risk factor reciprocity and symbiotic relationship between diabetes mellitus and periodontitis. Diabetes causes glycation of proteins; increases collagenase activity; impairs the function of neutrophils, lymphocytes, monocytes, and inflammatory response; causes periodontal vascular changes due to micro- and macroangiopathy and bone loss; and delays wound healing [2]. Periodontitis is a source of inflammatory mediators, released into the bloodstream and increasing insulin resistance [2]. Several common pathological pathways have been also described for diabetes mellitus and periodontitis, such as immune mechanisms, genetic component, as well as an improper response to environmental factors [2].

2.7 Periodontitis and aging

The population of periodontal bacteria increases with age [45]. Age is also a nonmodifiable cardiovascular risk factor. Several authors reported an increased arterial stiffness in patients with periodontitis [46–48]. Arterial stiffness indicates the impaired elastic vessel wall properties, is related to atherosclerosis and arteriosclerosis, and is considered a predictor of cardiovascular events [49].

2.8 Tooth loss

In adults, the most important causes of tooth loss are dental caries and, in patients older than 40 years, periodontitis. In patients with periodontitis, chronic inflammation is responsible for the destruction of the periodontal ligament and the resorption of the alveolar bone, enabling tooth loss [3]. Tooth loss, a marker of poor oral health, influences nutrient intake and food choices, with reduced intake of vegetables and dietary fibers, as well as socializing, and psychosocial risk factors, including social isolation, which are associated with a worse prognosis in cardiovascular disorders [50]. A healthy diet, rich in fruits and vegetables, is a cornerstone of cardiovascular disease prevention [4, 51]. Tooth loss is also associated with age, which is a cardiovascular risk factor. Observational studies linked tooth loss with coronary heart disease, heart failure, stroke, peripheral vascular disease, and cardiovascular mortality [30, 52]. The link tooth loss-cardiovascular disorders is explained by oral bacteria, especially *Streptococcus sanguis* and *Actinobacillus actinomycetemcomitans*, which contribute to a systemic inflammation with higher levels of C-reactive protein, endothelial dysfunction, atherosclerosis progression, and plaque instability [50]. On the other hand, a positive relationship was reported between tooth loss and all-cause mortality in persons who lost more than 10 teeth, but not for circulatory mortality, according to a meta-analysis including 18 prospective studies [50]. It must be mentioned that tooth extractions are not always disease related, because they may provide space for a full denture [30].

2.9 Periodontal interventions and cardiovascular outcomes

Periodontal interventions reduced cardiovascular risk, systemic inflammation, dyslipidemia, blood pressure, and endothelial dysfunction and may prevent embolic events and recurrence of arrhythmia [13, 43, 53]. Plaque removal in patients with stable coronary artery disease and periodontitis leads to lower levels of CRP, IL-6, and IL-8 [54]. Dental scaling (no more than twice per year) was associated with a lower risk of atrial fibrillation or flutter, probably due to its protective effect on PD [33]. Despite the association between periodontitis and atherosclerotic cardiovascular disease, periodontal interventions do not improve cardiovascular outcomes [30, 54]. Poor oral health can be considered as a risk marker for cardiovascular disorders, a biomarker of the severity of atherosclerosis [30].

3. Dental caries and lesions of endodontic origin

Dental caries and chronic apical periodontitis are different stages of the same inflammatory condition, and the focal infection theory has gained again attention [18, 21].

Streptococcus mutans, the bacteria commonly associated with dental caries, was identified also in the atherosclerotic plaque, suggesting a proatherogenic potential of dental caries, and may also cause bacterial endocarditis [21, 55]. It takes less than 1 minute after an oral intervention for the oral bacteria to reach the heart or the peripheral capillaries [21]. Normally, microorganisms are eliminated within minutes, but in patients with valvular heart disease, bacteremia may cause infective endocarditis [18].

A positive association between apical periodontitis and cardiovascular disease was also demonstrated [56]. Apical periodontitis, the inflammatory injury around the apex of a tooth due to gram-negative bacteria, may also cause systemic effects due to cytokines (interleukin 1 beta, interleukins 2, 6, 8, and 17, TNF alpha), reactive oxygen species, and matrix metalloproteinase, similar to chronic periodontitis [18, 21]. Interleukin 1 (IL-1) level, the predominant form of interleukin found in endodontic injuries, is involved in the formation, growth, and destabilization of the atherosclerotic plaque [18]. Interleukin 6 (IL-6) was associated with formation and activation of osteoclasts and also with unstable angina, left ventricular dysfunction, diabetes mellitus and its complications, hypertension, and obesity [18]. Interleukin 8 is associated with irreversible pulpitis and osteolysis in apical abscesses but also with angiogenesis and plaque formation [18]. TNF alpha was related to osteoclast activation and bone resorption, production of Il-6 and C-reactive protein, expression of cell adhesion molecules, smooth muscle cell proliferation, and lipid metabolism [18]. Interleukin 17 regulates matrix metalloproteinases, responsible for tissue destruction, and is involved in expression of genes encoding pro-inflammatory cytokines, endothelial damage, and cell apoptosis [18].

Periapical disease was associated with hypertension and stroke [57, 58]. Comparing aspects of periapical lesion formation in hypertensive and normotensive conditions using hypertensive and wild-type control mice, no differences were noticed in periapical lesion size and cytokines expressions, but hypertensive rats showed an elevated number of osteoclasts, responsible for bone destruction [59]. The link between bone destruction and hypertension is angiotensin II, able to upregulate RANKL expression in osteoblasts [59].

4. Sucrose

Sucrose enables cariogenesis in the presence of *Streptococcus mutans*. Dale et al. demonstrated an increased risk of congenital heart defects (patent ductus arteriosus, valvular pulmonary stenosis, ventricular septal defect, atrial septal defect) in offsprings of women after intake of sucrose-sweetened soft beverages during pregnancy [60].

Sucrose can also influence cardiovascular risk factors, especially overweight and obesity, diabetes mellitus, and blood lipids. The cardiovascular risk associated with a high sucrose intake is due to fructose-induced increase in blood triglycerides by stimulation of de novo lipogenesis and impairment of postprandial lipoprotein clearance [61]. The literature search did not provide a strong association between sucrose intake and LDL or HDL cholesterol level [61]. Only one study reported increase of LDL cholesterol and apoprotein B due to fructose consumption [62]. Daily fructose intake increasing blood triglycerides is of 50–60 g daily or above [61].

Intake of sweetened beverages, important contributors to free sugar intake and source of hidden calories, was associated with hypertension and impaired fasting glycemia in a study including 1158 young healthy participants [63]. Development of hypertension due to sucrose intake was related to slight insulin resistance with impaired nitric oxide synthase action, elevated lipoperoxidation, and decreased nonenzymatic antioxidant capacity and sirtuin 3 [64].

5. Fluorides

Fluorides enable synthesis of calcium fluoroapatite, which promotes remineralization of the enamel subjected to cariogenic factors and inhibits bacterial metabolism [2]. On the other hand, it has been demonstrated that fluoride induces damage to cardiomyocytes, due to Ca^{2+} metabolic disorder, and an abnormal expression of cardiac troponin T and I [65]. Fluoride can enter the cells; in excessive amounts it can cause serious damage to the cytoskeleton, nuclear condensation, myocardial fiber breakage, calcium overload, and mitochondrial dissolution, which impairs ATP production [65, 66]. The mentioned changes could explain myocardial ischemia, myocardial infarction, and heart failure associated with high fluoride intake, involving increased oxidative stress, apoptosis, and necrosis [67]. Fluoride in drinking water is nephrotoxic according to a study performed in rats with experimental chronic kidney disease, increasing the incipient aortic calcifications [68].

6. Biomarkers linking cardiovascular and oral disorders

Several serum biomarkers related especially to the peripheral inflammatory and immune response and oxidative stress link chronic periodontitis to atherosclerotic cardiovascular disease [13, 40]. The gingival index was associated with fibrinogen and white blood cell counts in periodontal patients and controls, adjusted for age, smoking, and socioeconomic status [69]. C-reactive protein (CRP) is a marker of both low-grade inflammations associated with periodontitis and apical lesions of endodontic origin and a risk indicator of cardiovascular events [13, 21, 70]. Apical lesions of endodontic origin were associated with the most promising biomarker of subclinical atherosclerosis: high-sensitivity C-reactive protein [70, 71].

Besides CRP, elevated cytokines were also associated with atrial fibrillation in patients with periodontitis [13]. IgA antibodies to periodontal pathogens were also revealed in patients with periodontitis [13].

Brain natriuretic peptides, released from ventricular cardiomyocytes in response to pressure and volume overload, were also increased in patients with periodontitis, related to severity of the disorder [3]. Ischemia-modified albumin, a marker of myocardial ischemia and end product of oxidative stress, was increased in patients with chronic periodontitis compared to periodontally healthy controls and decreased after nonsurgical periodontal therapy [72].

Matrix metalloproteinases, markers of plaque vulnerability, and subclinical atherosclerosis were also associated with periodontitis [12, 73].

7. Conclusions

It is important to recognize the importance of oral health, especially of chronic oral infections, for cardiovascular health and quality of life, considering that oral disorders are diagnosed easier and earlier than cardiovascular diseases. Several links were found between oral and cardiovascular disorders, including common risk factors, microbiological, clinical, inflammatory, and molecular markers. Oral bacteria, such as *Streptococcus mutans*, *Porphyromonas gingivalis*, *Aggregatibacter actinomycetemcomitans*, and *Prevotella intermedia*, may represent links between oral and cardiovascular disorders.

Both dentists and cardiologists, as well as other medical health-care providers, must extend their roles, considering the cross talk between oral and cardiovascular disorders. Markers of oral health enable screening of several cardiovascular disorders.

Conflict of interest

There is no conflict of interest.

Author details

Ioana Mozos[1,2]* and Dana Stoian[3]

1 Department of Functional Sciences, "Victor Babes" University of Medicine and Pharmacy, Timisoara, Romania

2 Center for Translational Research and Systems Medicine, "Victor Babes" University of Medicine and Pharmacy, Timisoara, Romania

3 2nd Department of Internal Medicine, "Victor Babes" University of Medicine and Pharmacy, Timisoara, Romania

*Address all correspondence to: ioanamozos@umft.ro

IntechOpen

References

[1] Sheiham A. Oral health, general health and quality of life. Bulletin of the World Health Organization. [Internet]. Available from: https://www.who.int/bulletin/volumes/83/9/editorial30905html/en/ [Accessed: 17 February 2019]

[2] Mozos I. Pathophysiology. In: Lecture Notes for Dental Medicine. Saarbrücken: Lambert Academic Publishing; 2015

[3] Mesa F, Magan-Fernandez A, Castellino G, et al. Periodontitis and mechanisms of cardiometabolic risk: Novel insights and future perspectives. Biochimica et Biophysica Acta, Molecular Basis of Disease. 2019;**1865**(2):476-484. DOI: 10.1016/j.bbadis.2018.12.001

[4] Piepoli MF, Hoes AW, Agewall S, et al. European guidelines on cardiovascular disease prevention in clinical practice. The sixth joint task force of the European society of cardiology and other societies on cardiovascular disease prevention in clinical practice (constituted by representatives of 10 societies and by invited experts) developed with the special contribution of the European association for cardiovascular prevention and rehabilitation (EACPR). European Heart Journal. 2016;**37**(29):2315-2381

[5] Meurman JH, Sanz M, Janket SJ, et al. Oral health, atherosclerosis, and cardiovascular disease. Critical Reviews in Oral Biology and Medicine. 2004;**15**:403-413

[6] Behle JH, Papapanou PN. Periodontal infections and atherosclerotic vascular disease: An update. International Dental Journal. 2006;**56**:256-262

[7] Blaizot A, Vergnes JN, Nuwwareh S, et al. Periodontal diseases and cardiovascular events: Meta-analysis of observational studies. International Dental Journal. 2009;**59**:197-209

[8] Kebschull M, Demmer RT, Papapanou PN. Gum bug, leave my heart alone!—Epidemiologic and mechanistic evidence linking periodontal infections and atherosclerosis. Journal of Dental Research. 2010;**89**:879-902

[9] Suzuki J, Aoyama N, Aoki M, et al. High incidence of periodontitis in Japanese patients with abdominal aortic aneurysm. International Heart Journal. 2014;**55**(3):268-270

[10] Hajishengallis G, Lamont RJ. Beyond the red complex and into more complexity: The polymicrobial synergy and dysbiosis (PSD) model of periodontal disease etiology. Molecular Oral Microbiology. 2012;**27**(6):409-419. DOI: 10.1111/j.2041-1014.2012.00663.x

[11] Kurihara N, Inoue Y, Iwai T, et al. Detection and localization of periodontopathic bacteria in abdominal aortic aneuysms. European Journal of Vascular and Endovascular Surgery. 2004;**28**:553-558

[12] Sekinishi A, Suzuki J, Aoyama N, et al. Periodontal pathogen *Aggregatibacter actinomycetemcomitans* deteriorates pressure overload-induced myocardial hypertrophy in mice. International Heart Journal. 2012;**53**:324-330

[13] Im SI, Heo J, Kim BJ, et al. Impact of periodontitis as representative of chronic inflammation on long-term clinical outcomes in patients with atrial fibrillation. Open Heart. 2018;**5**(1):e000708. DOI: 10.1136/openhrt-2017-000708

[14] Ettinger G, MacDonald K, Reid G, et al. The influence of the human microbiome and probiotics on

cardiovascular health. Gut Microbes. 2014;**5**(6):719-728

[15] Chistiakov DA, Orekhov AN, Bobryshev YV. Links between atherosclerotic and periodontal disease. Experimental and Molecular Pathology. 2016;**100**(1):220-235. DOI: 10.1016/j. yexmp.2016.01.006

[16] Mozos I, Malainer C, Horbańczuk J, et al. Inflammatory markers for arterial stiffness in cardiovascular diseases. Frontiers in Immunology. 2017;**8**:1058. DOI: 10.3389/fimmu.2017.01058

[17] Khlgatian M, Nassar H, Chou HH, et al. Fimbria-dependent activation of cell adhesion molecule expression in *Porphyromonas gingivalis*-infected endothelial cells. Infection and Immunity. 2002;**70**:257-267

[18] Garg P, Chaman C. Apical periodontitis—Is it accountable for cardiovascular diseases? Journal of Clinical and Diagnostic Research. 2016;**10**(8):ZE08-ZE12. DOI: 10.7860/ JCDR/2016/19863.8253

[19] Akhi R, Wang C, Nissinen AE, et al. Salivary IgA to MAA-LDL and oral pathogens are linked to coronary disease. Journal of Dental Research. 2019;**98**(3):296-303. DOI: 10.1177/0022034518818445

[20] Bronzato S, Durante A. Dietary supplements and cardiovascular diseases. International Journal of Preventive Medicine. 2018;**9**:80. DOI: 10.4103/ijpvm.IJPVM_179_17

[21] Bains R, Bains VK. Lesions of endodontic origin: An emerging risk factor for coronary heart disease. Indian Heart Journal. 2018;**70**(Suppl 3): S431-S434. DOI: 10.1016/j. ihj.2018.07.004

[22] Kure K, Sato H, Aoyama N, et al. Accelerated inflammation in peripheral artery disease patients with

periodontitis. Journal of Periodontal & Implant Science. 2018;**48**(6):337-346. DOI: 10.5051/jpis.2018.48.6.337

[23] Paraskevas KI, Mikhailidis DP, Giannoukas AD. Periodontitis and abdominal aortic aneurysms: A random association or a pathogenetic link? International Angiology. 2009;**28**(6):431-433

[24] Delbosc S, Alsac JM, Journe C, et al. *Porphyromonas gingivalis* participates in pathogenesis of human abdominal aortic aneurysm by neutrophil activation. Proof of concept in rats. PLoS One. 2011;**6**(4):e18679. DOI: 10.1371/journal.pone.0018679

[25] Suzuki J, Aoyama N, Aoki M, et al. Incidence of periodontitis in Japanese patients with cardiovascular diseases: A comparison between abdominal aortic aneurysm and arrhythmia. Heart and Vessels. 2015;**30**(4):498-502. DOI: 10.1007/s00380-014-0507-6

[26] Salhi L, Rompen E, Sakalihasan N, et al. Can periodontitis influence the progression of abdominal aortic aneurysm? A systematic review. Angiology. 30 Dec 2018:3319718821243. DOI: 10.1177/0003319718821243

[27] Figuero E, Lindahl C, Marín MJ, et al. Quantification of periodontal pathogens in vascular, blood, and subgingival samples from patients with peripheral arterial disease or abdominal aortic aneurysms. Journal of Periodontology. 2014;**85**(9):1182-1193. DOI: 10.1902/jop.2014.130604

[28] Aoyama N, Suzuki J, Ogawa M, et al. Toll-like receptor-2 plays a fundamental role in periodontal bacteria-accelerated abdominal aortic aneurysms. Circulation Journal. 2013;**77**:1565-1573

[29] Aoyama N, Suzuki JI, Ogawa M, et al. Clarithromycin suppresses the periodontal bacteria-accelerated

abdominal aortic aneurysms in mice. Journal of Periodontal Research. 2012;**47**:463-469

[30] Joshy G, Arora M, Korda RJ, et al. Is oral health a risk marker for incident cardiovascular disease hospitalisation and all-cause mortality? Findings from 172630 participants from the prospective 45 and Up Study. BMJ Open. 2016;**6**:e012386. DOI: 10.1136/bmjopen-2016-012386

[31] Hanatani T, Suzuki J, Ogawa M, et al. The periodontal pathogen *Aggregatibacter actinomycetemcomitans* deteriorates ventricular remodeling after myocardial infarction in mice. International Heart Journal. 2012;**53**:253-256

[32] Yu G, Yu Y, Li YN, et al. Effect of periodontitis on susceptibility to atrial fibrillation in an animal model. Journal of Electrocardiology. 2010;**43**(4):359-366

[33] Chen DY, Lin CH, Chen YM, et al. Risk of atrial fibrillation or flutter associated with periodontitis: A nationwide, population-based, cohort study. PLoS One. 2016;**11**(10):e0165601. DOI: 10.1371/journal.pone.0165601

[34] Aoyama N, Suzuki JI, Kobayashi N, et al. Detrimental effects of specific periodontopathic bacterial infection on tachyarrhythmia compared to bradyarrhythmia. BMC Cardiovascular Disorders. 2017;**17**(1):267. DOI: 10.1186/s12872-017-0703-2

[35] Nepomuceno R, Pigossi SC, Finoti LS, et al. Serum lipid levels in patients with periodontal disease: A meta-analysis and meta-regression. Journal of Clinical Periodontology. 2017;**44**(12):1192-1207. DOI: 10.1111/jcpe.12792

[36] Yamazaki K, Tabeta K, Nakajima T. Periodontitis as a risk factor for atherosclerosis. Journal of Oral Biosciences. 2011;**53**(3):221-232

[37] Sridhar R, Byakod G, Pudakalkatti P, et al. A study to evaluate the relationship between periodontitis, cardiovascular disease and serum lipid levels. International Journal of Dental Hygiene. 2009;**7**(2):144-150. DOI: 10.1111/j.1601-5037.2008.00318.x

[38] Saxlin T, Suominen-Taipale L, Kattainen A, et al. Association between serum lipid levels and periodontal infection. Journal of Clinical Periodontology. 2008;**35**(12):1040-1047. DOI: 10.1111/j.1600-051X.2008.01331.x

[39] Miyazawa H, Honda T, Miyauchi S, et al. Increased serum PCSK9 concentrations are associated with periodontal infection but do not correlate with LDL cholesterol concentration. Clinica Chimica Acta. 2012;**413**(1-2):154-159. DOI: 10.1016/j.cca.2011.09.023

[40] Tabeta K, Hosojima M, Nakajima M, et al. Increased serum PCSK9, a potential biomarker to screen for periodontitis, and decreased total bilirubin associated with probing depth in a Japanese community survey. Journal of Periodontal Research. 2018;**53**(3):446-456. DOI: 10.1111/jre.12533

[41] Rivas-Tumanyan S, Campos M, Zevallos JC. Periodontal disease, hypertension, and blood pressure among older adults in Puerto Rico. Journal of Periodontology. 2013;**84**(2):203-211. DOI: 10.1902/jop.2012.110748

[42] Pietropaoli D, Del Pinto R, Ferri C, et al. Poor oral health and blood pressure control among US hypertensive adults: Results from the national health and nutrition examination survey 2009 to 2014. Hypertension. 2018;**72**(6):1365-1373. DOI: 10.1161/HYPERTENSIONAHA.118.11528

[43] D'Aiuto F, Parkar M, Nibali L, et al. Periodontal infections cause changes in traditional and novel cardiovascular risk factors: Results from a randomized controlled clinical trial. American Heart Journal. 2006;**151**(5):977-984

[44] Holmlund A, Holm G, Lind L. Severity of periodontal disease and number of remaining teeth are related to the prevalence of myocardial infarction and hypertension in a study based on 4,254 subjects. Journal of Periodontology. 2006;**77**(7):1173-1178

[45] Vlachojannis C, Dye BA, Herrera-Abreu M, et al. Determinants of serum IgG responses to periodontal bacteria in a nationally representative sample of US adults. Journal of Clinical Periodontology. 2010;**37**(8):685-696. DOI: 10.1111/j.1600-051X.2010.01592.x

[46] Hayashida H, Saito T, Kawasaki K, et al. Association of periodontitis with carotid artery intima-media thickness and arterial stiffness in community-dwelling people in Japan: The Nagasaki Islands study. Atherosclerosis. 2013;**229**(1):186-191. DOI: 10.1016/j.atherosclerosis.2013.04.002

[47] Shanker J, Setty P, Arvind P, et al. Relationship between periodontal disease, *Porphyromonas gingivalis*, peripheral vascular resistance markers and coronary artery disease in Asian Indians. Thrombosis Research. 2013;**132**(1):e8-e14. DOI: 10.1016/j.thromres.2013.04.023

[48] Kapellas K, Jamieson LM, Do LG, et al. Associations between periodontal disease and cardiovascular surrogate measures among Indigenous Australians. International Journal of Cardiology. 2014;**173**(2):190-196. DOI: 10.1016/j.ijcard.2014.02.015

[49] Mozos I, Borzak G, Caraba A, et al. Arterial stiffness in hematologic malignancies. OncoTargets and Therapy. 2017;**10**:1381-1388. DOI: 10.2147/OTT.S126852

[50] Peng J, Song J, Han J, et al. The relationship between tooth loss and mortality from all causes, cardiovascular diseases, and coronary heart disease in the general population: Systematic review and dose-response metanalysis of prospective cohort studies. Bioscience Reports. 2019;**39**(1):BSR20181773. DOI: 10.1042/BSR20181773

[51] Mozos I, Stoian D, Caraba A, et al. Lycopene and vascular health. Frontiers in Pharmacology. 2018;**9**:521. DOI: 10.3389/fphar.2018.00521

[52] Watt RG, Tsakos G, de Oliveira C, et al. Tooth loss and cardiovascular disease mortality risk-results from the Scottish Health Survey. PLoS One. 2012;**7**:e30797

[53] Teeuw WJ, Slot DE, Susanto H, et al. Treatment of periodontitis improves the atherosclerotic profile: A systematic review and meta-analysis. Journal of Clinical Periodontology. 2014;**41**(1): 70-79. DOI: 10.1111/jcpe.12171

[54] Montenegro MM, Ribeiro IWJ, Kampits C, et al. Randomized controlled trial of the effect of periodontal treatment on cardiovascular risk biomarkers in patients with stable coronary artery disease: Preliminary findings of 3 months. Journal of Clinical Periodontology. 2019;**46**:321-331. DOI: 10.1111/jcpe.13085

[55] Kesavalu L, Lucas AR, Verma RK, et al. Increased atherogenesis during *Streptococcus mutans* infection in ApoE-null mice. Journal of Dental Research. 2012;**91**:255-260

[56] Berlin-Broner Y, Febbraio M, Levin L. Association between apical periodontitis and cardiovascular diseases: A systematic review of the literature. International Endodontic Journal. 2017;**50**(9):847-859. DOI: 10.1111/iej.12710

[57] Grau AJ, Buggle F, Ziegler C, et al. Association between acute

cerebrovascular ischemia and recurrent infection. Stroke. 1997;**28**:1724-1729

[58] Segura-Egea JJ, Jimenez-Moreno E, Calvo-Monroy C, et al. Hypertension and dental periapical condition. Journal of Endodontia. 2010;**36**:1800-1804

[59] Martins CM, Sasaki H, Hirai K. Relationship between hypertension and periapical lesion: An in vitro and in vivo study. Brazilian Oral Research. 2016;**30**(1):e78. DOI: 10.1590/1807-3107BOR-2016.vol30.0078

[60] Dale MTG, Magnus P, Leirgul E, et al. Intake of sucrose-sweetened soft beverages during pregnancy and risk of congenital heart defects (CHD) in offspring: A Norwegian pregnancy cohort study. European Journal of Epidemiology. 19 Jan 2019:1-14. DOI: 10.1007/s10654-019-00480-y

[61] Tappy L, Morio B, Azzout-Marniche D, et al. French recommendations for sugar intake in adults: A novel approach chosen by ANSES. Nutrients. 2018;**10**(8):989. DOI: 10.3390/nu10080989

[62] Stanhope KL, Bremer AA, Medici V, et al. Consumption of fructose and high fructose corn syrup increase postprandial triglycerides, LDL-cholesterol, and apolipoprotein-B in young men and women. The Journal of Clinical Endocrinology & Metabolism. 2011;**96**:E1596-E1605

[63] Popa AR, Vesa CM, Uivarosan D, et al. Cross sectional study regarding the association between sweetened bevarages intake, fast-food products, body mass index, fasting blood glucose and blood pressure in the young adults from north-western Romania. Revista de Chimie. 2019;**70**(1):156-160

[64] Castrejón-Téllez V, Villegas-Romero M, Pérez-Torres I, et al. Effect of sucrose ingestion at the end of a critical window that

increases hypertension susceptibility on peripheral mechanisms regulating blood pressure in rats. Role of sirtuins 1 and 3. Nutrients. 2019;**11**(2):E309. DOI: 10.3390/nu11020309

[65] Wang HW, Liu J, Zhao J, et al. Ca^{2+} metabolic disorder and abnormal expression of cardiac troponin involved in fluoride-induced cardiomyocyte damage. Chemosphere. 2018;**201**:564-570. DOI: 10.1016/j.chemosphere.2018.03.053

[66] Panneerselvam L, Raghunath A. Perumal: Acute fluoride poisoning alters myocardial cytoskeletal and AMP signaling proteins in rats. International Journal of Cardiology. 2017;**229**:96-101

[67] Panneerselvam L, Govindarajan V, Ameeramja J, et al. Single oral acute fluoride exposure causes changes in cardiac expression of oxidant and antioxidant enzymes, apoptotic and necrotic markers in male rats. Biochimie. 2015;**119**:27-35. DOI: 10.1016/j.biochi.2015.10.002

[68] Martín-Pardillos A, Sosa C, Millán Á. Effect of water fluoridation on the development of medial vascular calcification in uremic rats. Toxicology. 2014;**318**:40-50. DOI: 10.1016/j.tox.2014.01.012

[69] Beck JD, Offenbacher S, Williams R, et al. Periodontitis: A risk factor for coronary heart disease? Annals of Periodontology. 1998;**3**(1):127-141

[70] Garrido M, Cárdenas AM, Astorga J, et al. Elevated systemic inflammatory burden and cardiovascular risk in young adults with endodontic apical lesions. Journal of Endodontia. 2019;**45**(2):111-115. DOI: 10.1016/j.joen.2018.11.014

[71] Tibaut M, Caprnda M, Kubatka P, et al. Markers of atherosclerosis: Part 1—Serological markers. Heart, Lung & Circulation. 4 Oct 2018.

pii: S1443-9506(18)31916-4. DOI:
10.1016/j.hlc.2018.06.1057

[72] Tayman MA, Önder C, Kurgan Ş,
et al. A novel systemic indicator of
periodontal tissue damage: Ischemia
modified albumin. Combinatorial
Chemistry & High Throughput
Screening. 2018;**21**(8):544-549. DOI:
10.2174/1386207321666181018165255

[73] Tibaut M, Caprnda M, Kubatka
P, et al. Markers of atherosclerosis:
Part 2—Genetic and Imaging Markers.
Heart, Lung & Circulation. 29 Sep
2018. pii: S1443-9506(18)31914-0. DOI:
10.1016/j.hlc.2018.09.006

Chapter 4

The "Weight" of Obesity on Arterial Hypertension

Annalisa Noce and Nicola Di Daniele

Abstract

The prevalence of obesity and its related diseases are increasing worldwide. This phenomenon has been observed not only in adults but also in adolescents and children. Numerous scientific studies have revealed a direct correlation between the increase in blood pressure and weight gain. In fact, visceral fat can contribute to the rise in blood pressure because it is associated with an increased production of inflammatory cytokines (such as interleukin-1-β, tumor necrosis factor-α and interleukin-6) and inflammatory factors (such as C-reactive protein), inducing endothelial dysfunction and consequently arterial hypertension (AH). Insulin resistance, which develops in obese individuals, may represent an additional risk factor in the onset of AH. Postprandial hyperglycemia is not able to inhibit lipolysis, inducing a greater release of free fatty acids causing metabolic abnormalities, oxidative stress and vascular dysfunction. In this chapter, we will examine the mechanisms that correlate obesity to hypertension, such as the involvement of the sympathetic nervous system, metabolic and renal alterations. Finally, the pharmacological and nutritional treatment of obesity-related hypertension will be described.

Keywords: obesity, metabolic syndrome, renin-angiotensin system, obesity-hypertension link, anti-obesity drugs

1. Introduction

The World Health Organization (WHO) defines obesity as the clinical condition in which a subject presents a body mass index (BMI) ≥ 30 kg/m^2. This condition can be further classified in three stages: stage 1 BMI 30-34.99 kg/m^2, stage 2 BMI 35–39.99 kg/m^2 and stage 3 BMI ≥ 40 kg/m^2 [1].

Obesity and being overweight (BMI ≥ 25 kg/m^2) [2] have become in recent years a substantial health burden due to their growing prevalence.

In the USA, the prevalence of obesity in adults has increased by 39.6% from 2015 to 2016, and it is forecasted to reach the astounding number of 2.1 billion people by 2030 [3]. In China, the percentage of obese men and women in 2013 was, respectively, of 3.8 and 5.8%, while in Japan it was of 4.5 and 3.3%. In Eastern Europe during 2013, the percentage of obese adult subjects did not differ much from that in the USA: 21% of the adult population resulted obese, with equal gender distribution [2]. In Germany, a recent study has highlighted that 35.4% of the adult population was overweight and the 21.3% was obese, in Italy 34.9% was overweight and 12.3% obese [4].

All epidemiologic studies conducted till date have confirmed that the global prevalence of obesity is constantly rising [5]. Therefore, it is becoming a sanitary emergency, both in terms of human resources and economically. In fact, it is worthy to consider that excess of adipose tissue is associated with an increase in cardiovascular (CV) risk and with the precocious insurgence of CV diseases [6]. It has been underlined that obesity is characterized by an augmented activation both the sympathetic nervous system (SNS) [7], and of renin-angiotensin-aldosterone system (RAAS), which play a fundamental role in the physiopathology of AH [8]. It is estimated that about 75% of hypertension incidence is directly correlated to the contextual presence of obesity, characterizing the form of obesity-related hypertension [9, 10].

2. Physiopathological mechanisms of obesity-related hypertension

The association between body weight and arterial pressure was made for the first time during the 1960s, in the Framingham Heart Study [11]. However, the nature of such correlation remained unknown until the latter half of the 1980s, when a series of studies highlighted the possible mechanisms, which correlated these two clinical entities [12–14]. Such studies took inspiration from the clinical observation made by Vague [15], who observed that metabolic and CV complications linked to obesity were more frequent in subjects with a phenotype of "android" obesity (most fat localized in the upper part of the body) that with those with a "gynoid" phenotype (most fat localized in the inferior part of the body). Successively, during the 1980s, other population studies have been conducted which utilized the waist/ hip ratio as a quantitative index of the visceral fat, demonstrating that a higher ratio was correlated with a significant increase of CV risk [12–14]. Further studies shed light on the presence of insulin resistance in the android phenotype [16, 17]. Furthermore, the possible association between insulin resistance and the presence of AH was evaluated in both obese and non-obese subjects, constructing the basis for the comprehension of the physiopathological mechanisms of obesity-related hypertension [18, 19].

The accumulation of excess adipose tissue triggers a cascade of events, which induce a rise in blood pressure values in both children and adults [20, 21]. The physiopathological mechanism at the basis of the insurgence of hypertension is complex, and encompasses the following: activation of the SNS through the action of hyperleptinemia and hyperinsulinemia, vascular damage caused by a chronic low-grade inflammatory state, endothelial dysfunction, oxidative stress and finally vasoconstriction coupled with fluid retention modulated by the RAAS activation [22–24].

2.1 Increased SNS activity

In the condition of obesity-related hypertension, hyperactivation of the SNS can be observed [25]. In such mechanism, total body fat distribution plays a pivotal role, as microneurography studies have demonstrated that the grade of SNS activity is greater in subjects who present visceral fat distribution [26, 27]. Moreover, it has been shown that there is a direct correlation between SNS activation and the waist/ hip ratio [28].

Numerous studies have demonstrated that obesity induces the alteration of the arterial baroreceptor control of sympathetic activity, which involves inhibitory and excitatory components [25].

Commonly, with obesity, a reduction of the parasympathetic tone can be observed with an increase of sympathetic activity with a reduction of heart rate variability [29, 30]. On the contrary, with weight loss, the parasympathetic tone and the heart rate variability increase.

Obesity is also associated to tissue SNS activation, at the levels of the heart, kidney and musculoskeletal tissue [31, 32]. Specifically, obese subjects present an increase of sympathetic renal nervous tissue activity, diagnosed by an increased concentration of renal norepinephrine [33]. Moreover, obese subjects with normal blood pressure have a suppressed cardiac SNS activity, while obese subjects with AH have an increase in cardiac SNS activity [33]. Consequently, it is hypothesized that the development of obesity-related hypertension plays a fundamental role in augmenting renal and cardiac sympathetic activity. In order to confirm such hypothesis, it is worth noting that renal denervation induces a reduction of blood pressure values and an increased sodium excretion in a canine model fed with a high fat diet [34].

Other mechanisms that would seem to be involved in the regulation of the activation of the SNS, determining other direct effects on CV homeostasis are hormonal, metabolic, inflammatory and endothelial factors. The possible relation between insulin and arterial blood pressure was initially quite controversial, however, recent studies have highlighted a possible role for such hormone in the physiopathological mechanism of obesity-related hypertension [35]. Since insulin stimulates the SNS, and obese subjects have an increase in SNS activity, it is hypothesized that the stimulation of the SNS is also mediated by insulin [36]. Such hypothesis would also explain the physiopathological mechanism at the basis of the elevated blood pressure values that are recorded in central obesity. This would also seem to be supported by some studies that have shown a concomitant reduction of arterial blood pressure and SNS activity in obese subjects who underwent insulin level reduction thanks to a low calorie diet [37]. Moreover, chronic hyperinsulinemia needs to be also correlated to an arterial dysfunction, which favors a mechanism of vasoconstriction. Insulin can exert a direct action on the kidney, by stimulating sodium reabsorption and consequently inducing sodium retention via direct interaction with renal tubules [38]. Therefore, obesity-induced hyperinsulinemia seems to contribute to the increase in arterial blood pressure values by acting on sodium retention, and the expansion of extracellular volume.

Another factor involved in SNS stimulation is leptin. This adipokine is produced by adipocytes, and its plasmatic concentration is directly correlated to the amount of fat mass of the subject [39]. Leptin induces appetite suppression and stimulates the SNS [40]. With regards to leptin concentration, gender variation has been observed: female individuals present higher hormonal levels and a greater receptor expression (ObR) compared to male subjects [41]. A possible explanation for such phenomenon has highlighted that subcutaneous adipose tissue, predominating in the female gender, produces a greater quantity of leptin compared to visceral adipose tissue [42].

Studies conducted on animal models have demonstrated that leptin infusion induces an increase in blood pressure values and SNS hyperactivation [43, 44].

From recent studies, it has emerged that leptin-mediated SNS stimulation could be seen as a mechanism to stabilize bodily weight and restore energetic equilibrium in obese patients, increasing energetic expenditure by stimulating brown fat thermogenesis [45].

2.2 Increased RAAS activity

Numerous studies have highlighted that urinary and plasmatic concentration of aldosterone is increased in obese subjects compared to normal weight subjects [46]. In particular, its plasmatic concentration results directly correlated with the quantity of visceral adipose tissue [47]. Various authors have shown how adipose tissue releases adipokines, which stimulate the adrenal glands to produce aldosterone, independently from the plasmatic activity of renin [48–50]. Therefore, RAAS activation is directly involved in the development of obesity-related hypertension. Obese subjects, especially if they present a substantial visceral fat quota, often have increased plasmatic renin activity, together with an enhancement in angiotensin converting enzyme (ACE), a greater concentration of aldosterone, angiotensinogen and angiotensin II [51]. In obese patients, RAAS activation is determined by a number of factors, some of which are constituted by physical renal compression induced by an increment in visceral fat, SNS hyperactivation and local activation of RAAS in the adipose tissue [52]. Adipose tissue, other than containing all RAAS components, is able to produce angiotensin II [53]. Even if in obese subjects the principal bulk of angiotensinogen continues to be produced by the liver (as in healthy subjects), it has been demonstrated that in obese patients, there is an increase of angiotensinogen produced by the adipose tissue [54]. To support this finding, an interesting study conducted on adipocyte-angiotensinogen deficient mice (Agt^{aP2}) has demonstrated that a fat-rich diet induces blood pressure increase in wild type rats, but causes no pressure increase in Agt^{aP2} rats, even if both groups present an equal increase in weight and fat mass [55]. Moreover, it is worth to point out that RAAS of adipose tissue not only produces angiotensin II, through ACE enzymatic activity, but also uses a less common mechanism that relies on the enzymatic activity of cathepsins and chymases [56].

To prove the hypothesis that RAAS activity has a significant role in the pathogenesis of obesity-related hypertension; a study has been conducted on obese subjects who have undergone bariatric surgery, which demonstrated a significant reduction in RAAS activity due to considerable weight loss [57].

A fundamental role in obesity-related hypertension is also carried out by aldosterone, as obese subjects present elevated levels of this hormone [58, 59], and since weight loss is not only associated with reduced plasmatic renin activity but also with aldosterone, independently from sodium intake. As a final analysis, weight loss also induces a beneficial effect on blood pressure values [58].

Aldosterone increase has also been observed in obese adolescents [59] and obese menopausal women [60]. Moreover, one should consider that the highest levels of plasmatic renin activity, aldosterone and ACE have been highlighted in subjects with visceral obesity but not in subjects with peripheral obesity [61]. Goodfriend et al. have confirmed that in subject affected by visceral obesity, adipose tissue is involved in the excess production of aldosterone, through the action of aldosterone releasing factors [62].

2.3 Changes in kidney function and hemodynamic in obesity-related hypertension

Obese subjects present an elevated risk to develop chronic kidney disease [63]. During obesity, an expansion of extracellular volume and an increase in blood flow in many tissues that leads to increased cardiac output, can be observed [64]. The latter, increases in a consistent manner with weight gain, and part of such increase is closely correlated to the blood flow needed to supply the excess of adipose tissue. Such blood flow increase is appreciable not only at the adipose tissue level, but also in other organs and tissues such as the heart, the kidney, the gastro-intestinal apparatus and the muscle [21, 65]. Excess blood flow, which can be observed at the level of other organs, is caused by the hypertrophy that such organs undergo because of obesity; it is secondary to increased metabolic demand and to the greater work load present in this pathological condition [66]. Renally, this translates into glomerular hyperfiltration that can be encountered in the initial phases of the pathology, which will successively progress in chronic kidney disease with reduced glomerular filtration [67]. During the initial stages of obesity, there is an augmented sodium tubular reabsorption with a consequent increase in sodium retention. In order to compensate such mechanism, the kidney undergoes vasodilation with subsequent hyperfiltration and increased filtration of water and electrolytes. However, this compensatory mechanism is incomplete and induces an extracellular volume expansion with an increase in blood pressure values. Therefore, obesity induces an increase sodium tubular reabsorption in the kidney via different mechanisms such as neural, hormonal and reno-vascular. The first involving the SNS, the second insulin and aldosterone, and the third angiotensin II [68]. During obesity, even with the expansion of extracellular volume, renin secretion by the kidneys still occurs. This is due to the action exerted by fat accumulated in the renal medulla and in the peri-renal adipose tissue [10, 69].

2.4 Inflammation and obesity-related hypertension

In obese subjects, adipose tissue dysfunction can be observed. It is characterized by a reduction in protective factor concentration such as adiponectin, nitric oxide and prostaglandins, and an increased release of pro-inflammatory adipokines such as resistin, leptin and visfatin, with subsequent development of low-grade inflammation. Cumulatively, this induces a metabolic and vascular dysfunction in the obese subjects [70, 71].

In pathologic conditions, adipocytes produce both inflammatory cytokines and extracellular matrix proteins, favoring the infiltration of immune cells in the adipose tissue and consequent inflammation [72]. In turn, the same infiltrative immune cells activate and release cytokines that can directly influence the adipocyte function or induce the secretion of pro-inflammatory adipokines. These effects are also evident at the level of the perivascular adipose tissue, particularly when adjacent to atherosclerotic or dysfunctional vessels in hypertensive subjects. In fact, in AH, perivascular adipose tissue inflammation can be observed. This kind of inflammation, is in turn involved in the dysfunction of vessels [73], favoring vasoconstriction and inhibiting endothelium-dependent vasodilation [72]. Such functional changes will have correspondent morphological changes: perivascular adipose tissue becomes pro-inflammatory, dedifferentiated and metabolically active. This tissue will produce a greater number of chemokines, such as RANTES, involved in the activation of monocyte/macrophages and CD8$^+$ T cells. Moreover, an increase in sympathetic innervation can be observed at the level of the perivascular adipose tissue, which is also involved in the mechanism of obesity-related hypertension [74].

Hypertensive subjects present at the perivascular adipose level an increment in T lymphocytes, antigen-presenting cells and factors involved in endothelial dysfunction. These factors explain the persistent relationship between hypertension and the atherosclerotic process [75]. During AH, one can observe in the perivascular adipose tissue an increase in both CD4 and CD8 and the expression of pro-inflammatory cytokines (TNF-α and INF-γ) [73, 76]. The pro-inflammatory cytokines modulate smooth muscle cell contraction, their migration and proliferation [77]. However, it is also worth considering that leptin is structurally similar to IL-6, IL-12 and IL-15 and is capable to induce leukocyte activation, chemotaxis, free radical production and the expression of endothelial adhesion molecules at the level of vascular smooth vessel cells. Moreover, pro-inflammatory cytokines (IL-17A and TNF-α) induce in the adipose tissue an increased production of leptin and resistin, which in turn cause an augmented expression of VCAM-1 and ICAM-1, causing vascular dysfunction and oxidative stress [78].

2.5 Obstructive sleep apnea

Obstructive sleep apnea (OSA) represents together with obesity, an independent risk factor for the development of AH.

The first study, which highlighted the presence of fluctuations in arterial blood pressure in course of complete or partial OSA, was conducted in 1972 [79]. OSA is characterized by a series of events that induce the collapse of the superior airways during sleep with consequent intermittent hypoxia, hypercapnia, negative intrathoracic pressure and increased activation of the SNS [80, 81]. Therefore, the intermittent hypoxia observed during OSA (which also activates the SNS) contributes to the development of obesity-related hypertension [82]. OSA-induced SNS activation is caused by a dual mechanism: on one hand from the stimulation of peripheral chemoreceptors, and on the other by the formation of ROS that contribute to systemic inflammation and endothelial dysfunction [83].

During hypoxia, it has been demonstrated that endothelin, which has a vasoconstrictor action is released [84], with the improvement of oxygenation instead, there is a decreased production of endothelin and consequent vasodilation. Phillips et al. have suggested that the cyclical alterations of endothelin production in OSA patients can contribute to the insurgence of AH [85]. Such clinical observations have been confirmed on OSA murine models [86]. Moreover, OSA patients present a higher incidence of non-dipping of nocturnal systolic pressure, indicator of an increased adrenergic tone [87].

3. Therapeutic approaches to the obesity-related hypertension

According to the 8th report of the Joint National Committee (JNC8), normal blood pressure systolic values are inferior to 130 mmHg and normal diastolic values are inferior to 80 mmHg [88].

In order to have reliable measurements, ESC/ESH recommends to perform three of them in an ambulatories setting, with intervals of 1 or 2 min between each reading. Moreover, it is advised to perform additional measurements, if the first one differ more than 10 mmHg between one another. The blood pressure value that will be given by an average of the last two readings [89].

Even if the definition and the categorization of AH have varied over time, there is a consensus that is persistent. The therapeutic target is fixed around values equal or inferior to 130/80 mmHg [89, 90].

The therapeutic approach of obesity-related hypertension is based on a pharmacological therapy (anti-hypertensive and anti-obesity pharmaceuticals) associated to a nutritional/compartmental intervention (**Figure 1**). A healthy lifestyle is a valid support to pharmacological therapy and allows the correction of certain deleterious habits such as physical activity, hypercaloric diet, sodium-rich diet and alcohol abuse.

3.1 Anti-hypertensive drugs

3.1.1 RAAS inhibitors

Numerous studies have suggested diuretics is RAAS antagonists are particularly effective in obese subjects [91, 92]. In fact, as previously described, since angiotensin is overexpressed in obesity and given its role in the development of obesity-related hypertension, ACE-inhibitors and angiotensin II receptor blockers (ARBs) are considered a valid therapeutic approach in such patients.

Moreover, when comparing ACE-inhibitors and ARBs to β-blockers and thiazide diuretics, it is apparent that RAAS antagonists are seldom associated to new cases of diabetes and induce less insulin resistance [93, 94]. Furthermore, ACE-inhibitors and ARBs do not appear correlated with weight gain, and carry out a nephroprotective

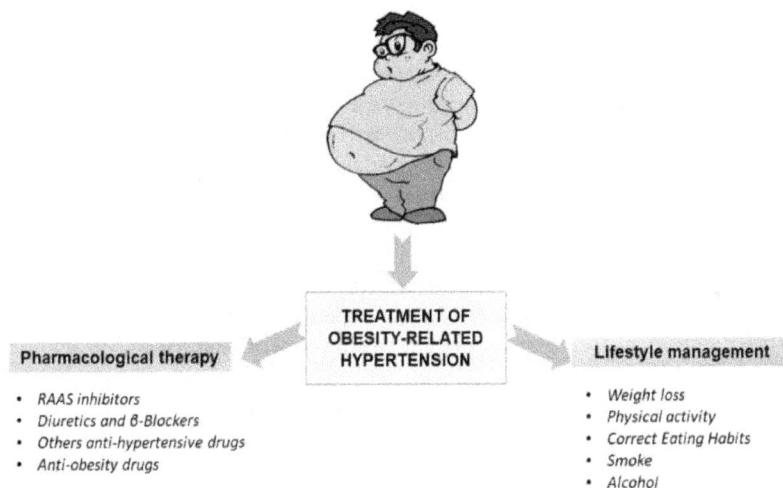

Figure 1.
Therapeutic approach of obesity-related hypertension.

action in diabetic patients, which is a frequent comorbidity in obese subjects. These pharmaceuticals also induce a 30% reduction of left ventricular hypertrophy, which has a high prevalence in subjects affected by obesity-related hypertension [95, 96].

In support of the above statements, a study conducted on 6083 hypertensive subjects (65–84 years of age) with an average BMI of 27.4 kg/m^2, has highlighted that starting the anti-hypertensive treatment with ACE-inhibitors will lead to a better outcome compared to an intervention with diuretics, independently of the improvement in blood pressure values [97].

3.1.2 Diuretics and β-blockers

Even if treatment with thiazide diuretics is frequently recommended in patients with AH [98], it must be kept in consideration that they present some dose-correlated collateral effects such as insulin resistance, hyperuricemia and dyslipidemia. Moreover, in obese patients who have a predisposition for type 2 diabetes mellitus and metabolic syndrome, this kind of pharmaceutical approach is harmful and should be avoided [8].

To demonstrate this, an interesting open label, randomized study conducted on CV disease free and non-diabetic, hypertensive patients had shown that the adverse metabolic effects related to the treatment with thiazide diuretics and β-blockers were more frequent in subjects with abdominal obesity. Such adverse metabolic effects manifested after only 9 weeks from the start of the treatment [99].

Therefore, according to guidelines such pharmaceuticals, should be used with caution in patients who are at risk of developing metabolic syndrome or altered fasting glucose, in order to avoid the insurgence of diabetes and other long-term complications [100, 101].

It is recommended to use a low dose of thiazides in case of need, in association with a careful monitoring of the lipid and glycemic profile. β-Blockers, in addition to inducing insulin resistance, are associated to body weight gain, because they are thought to reduce thermogenesis induced by diet and velocity of fat oxidation [102, 103]. In obese patients, the use of β-blockers should be limited to subjects with a precise CV indication (namely, cardiac insufficiency and previous myocardial infarction).

3.1.3 Other antihypertensive drugs

Calcium antagonists are antihypertensive drugs that induce diuresis and natriuresis, without exerting any effect on glucose and lipid metabolism.

They are constituted by two subclasses, such as non-dihydropyridines and dihydropyridines, with notably different pharmacological effects. The first are normally used for the treatment of cardiac arrhythmia and seem to have an anti-proteinuric effect, similar to that induced by ACE inhibitors [104–107], and seem to slow down the progression of diabetic nephropathy [108].

On the other hand, dihydropyridines seem to heighten albuminuria, a factor correlated to an increased CV risk [109]. Such increase seems to be caused by a dilation of the preglomerular afferent arteriole with a consequent increase of intraglomerular pressure [110].

An analysis by Avoiding Cardiovascular Events through Combination Therapy in Patients Living with Systolic Hypertension (ACCOMPLISH) has evaluated which type of anti-hypertensive treatment could have an impact on the CV outcome of patient based on their body surface area (evaluated thanks to their BMI). It has been highlighted that, while thiazide treatment induces minor CV protection in normal subjects compared to obese ones, patients treated with calcium antagonist (amlodipine) did not present differences in CV protection according to their BMI. Therefore, calcium antagonists appear better at exerting CV protecting action on hypertensive non-obese subjects [111].

Another clinical study has showed important differences on the CV mortality of patients treated with calcium antagonists, β-blockers or sartans, highlighting that patients treated with amlodipine were majorly protected from CV events and had a reduced risk of developing diabetes compared to those treated with β-blockers (atenolol) [112]. This study did not examine the impact induced by the body surface area on CV protection. Therefore, calcium antagonists emerge as a pharmaceutical alternative in the treatment of obesity-related hypertension with potential benefits [113], even if more clinical randomized trials are required to fully understand their beneficial effects [114, 115].

A retrospective observational study conducted in southern Italy, has examined what type of pharmaceuticals are used for the treatment of obesity-related hypertension in a clinical practice. It has highlighted that the clinicians do not differentiate between pharmaceuticals to use in relationship to the grade of obesity or the presence of metabolic syndrome. Moreover, it has been observed that antihypertensive pharmaceuticals, which cause negative effects on weight and metabolic profile, are still largely used in subjects at risk [116]. Such study underlines the necessity to conduct clinical randomized trials on a large population of patients affected by obesity-related hypertension, in order to standardize pharmacological antihypertensive therapy based on obesity severity, CV protection and limitation of metabolic complications.

3.2 Anti-obesity drugs

Lifestyle changes perform a key role in the treatment of obesity-related hypertension. According to the Obesity Education Initiative Working Group guidelines, anti-obesity drugs should represent a support strategy in the treatment of subjects with BMI ≥ 30 kg/m^2 without associated comorbidities, and in subjects with BMI ≥ 27 kg/m^2 with comorbidities [117]. Drugs used in the treatment for obesity can be divided into the following categories: (1) inhibitors of nutrient absorption (orlistat and acarbose); (2) appetite suppressors (phentermine and lorcaserin); (3) drugs used in the treatment for diabetes that determine weight loss (metformin and incretin therapy: GLP1 agonists and DPP-4 inhibitors) [8, 113].

3.2.1 Inhibitors of nutrient absorption

Orlistat is a gastrointestinal lipase inhibitor that causes a consequent reduction in absorption of dietary fat. Its use in clinical practice is limited by gastrointestinal adverse effects, which occur especially if patients have a high fat diet [118]. A study aimed at evaluating orlistat and sibutramine (serotonin and norepinephrine re-uptake inhibitor acting as an appetite suppressor, removed from the market in 2010 because of associated increased CV risk) [119] has highlighted the same level of efficacy in reducing BMI, body weight and waist circumference. Moreover, orlistat-treated subjects presented a significant reduction in blood pressure values, while they remained stable in sibutramine-treated subjects [120].

In order to reduce gastrointestinal adverse effects related to orlistat ingestion, a half dose pill (60 mg) has been produced that is still able to reduce fat absorption of 25% [121, 122].

A study has evaluated as secondary outcome, long-term effects related to orlistat treatment, demonstrating that about two-third of the weight loss was maintained over a 2 years compared to the placebo-treated group. The pharmaceutical intervention was combined during the first year with a hypocaloric diet, and during the second year with a weight maintaining diet. Moreover, patients treated with orlistat full dose (120 mg), presented after 2 years an improvement in systolic pressure values [123].

Acarbose is an oral antidiabetic. It inhibits intestinal alpha-glucosidase and pancreatic alpha-amylase. By inhibiting these enzymes, such drug impedes the digestion and absorption of complex sugars. It has been highlighted that acarbose induces a modest reduction in body weight [124]. A study performed on 110 obese subjects with BMI between 32 and 38 kg/m^2, who underwent a hypocaloric diet for 10–16 weeks and substantial weight loss, has highlighted that acarbose treatment does not induce significant effects on the stabilization of weight loss [125], confirming the modest effect of the drugs in body weight reduction.

3.2.2 Appetite suppressors

Appetite control represents a cardinal step in the treatment of obesity. Studies finalized to evaluate appetite control have demonstrated the existence of an endogenous system, which is able to stimulate, through orexigenic substances, and inhibit, through anorexigenic substances, food intake [126, 127]. Leptin and serotonin are two endogenous ligands which act contemporaneously inhibiting the hypothalamic feeding center, and stimulating the satiety center [128]. An important appetite suppressor to cite is fenfluramine, which was successively removed from the market because of its severe collateral effects. It acted on serotonin release at the level of the hypothalamus and induced significant weight loss [129].

In 2012, the Food Drug Administration (FDA) has approved the use of lorcaserin for the treatment of obesity [130]. This drug is a serotonin 2C receptor agonist (5-HT_{2c}), and appears to be efficacious in co-adjuvating weight loss in obese and overweight subjects in association with a hypocaloric diet and increased physical activity [131].

Lorcaserin, being selective for 5-HT_{2c} receptors, represents a more efficacious and safe drug, compared to other non-selective serotoninergic appetite suppressor drugs because it does not induce CV collateral effects [132].

Another drug belonging to the class of the appetite suppressors is phentermine, an adrenergic agonist that, thanks to the central nervous system and SNS activation, is capable to determine reduced food intake and increase basal metabolism [133]. The FDA approves phentermine use to treat obesity for a period no greater than 3 months. Since it causes an increase in the release of norepinephrine, it could lead to an increase in blood pressure values and cardiac frequency [134].

3.2.3 Drugs used in the treatment for diabetes that determine weight loss

Drugs used for the treatment of diabetes commonly determine weight loss and reduced fat accumulation. For overweight and obese subjects suffering of diabetes, the FDA has approved the use of hypoglycemic drugs, which are associated to weight loss and blood pressure reduction. Even if the effect is modest, it is however to be considered beneficial given the weight gain that is frequently associated with insulin and insulin secretagogue analogs.

Metformin is an oral hypoglycemic drug, used in type II diabetes treatment, that is capable of inducing modest weight loss as a consequence of a reduced hepatic production and intestinal absorption of glucose, and through the improvement of insulin sensitivity [135]. Such drug, as demonstrated by the Diabetes Prevention Program trial [136], is efficacious in reducing body weight in a follow-up period of 2.8 years in overweight diabetic subjects. However, metformin does not appear useful in reducing blood pressure, as was demonstrated in a series of clinical trials in which lifestyle modification resulted notably more efficacious in the control of blood pressure compared to this type drug [137].

Incretins are intestinal hormones secreted by enteroendocrine cells in the circulatory stream, few minutes after feeding. They regulate the quantity of post-feeding insulin secretion. There are two endogenous incretins such as glucose-dependent insulinotropic peptide (GIP) and glucagon-like peptide 1 (GLP-1); both are rapidly metabolized an enzyme called dipeptidyl peptidase-4 (DPP-4). These hormones, increasing insulin release by β pancreatic cells and inhibiting glucagon release, play an important role in glucose homeostasis. GLP-1 is the most abundant, but cannot be used for therapeutic scopes given its rapid degradation by DPP-4.

In 2005, exenatide was released, an injectable GLP-1 agonist, which mimics endogenous GLP-1 but has a prolonged action. Exenatide increases glucose-dependent insulin secretion, suppresses glucagon secretion and slows down gastric filling. Therefore, it is hypothesized that exenatide could have a role in the treatment for obesity, since it induced a sense of satiety and reduced food intake.

DPP-4 inhibitors (sitagliptin, saxagliptin and linagliptin) are pharmaceuticals which increase endogenous plasmatic levels of active incretins, prolonging their action [138]. Moreover, they also reduce the degradation of many vasoactive peptides. Studies on animal models of ischemia/reperfusion have demonstrated a positive effect of DPP-4 inhibitors. Endogenous GLP-1 exerts protective effects on the myocardium and has a vasodilatory action [139]. A study that compared different pharmaceutical therapies to tackle type II diabetes mellitus, such as exenatide and sitagliptin, has demonstrated that incretin treatment was associated with superior weight loss compared to the insulin-treated group. Weight loss was associated to a statistically significant systolic and diastolic blood pressure reduction in all the treated groups [140].

4. Life style management

The most common consequences related to obesity are the insurgence of essential AH, diabetes mellitus, chronic kidney disease, metabolic diseases, etc. [5, 141, 142]. Even if antihypertensive drugs are of primary importance in the treatment of AH, they should always be associated with healthy eating habits, adequate levels of physical activity and a correct lifestyle [143].

For this reason, it would be advised that hypertensive-obese patients follow nutritional counseling in order to evaluate their food habits and levels of physical activity [144]. In fact, incorrect food habits, scarce levels of physical activity and psychological factors such as depression can contribute to weight gain [145].

Nutritional intervention finalized to achieving weight loss in hypertensive-obese patients, should be personalized. Different types of diet exist, for example, strongly hypocaloric diet, balanced slightly hypocaloric diet, low-sodium diet, hypolipidic diet, hypoglucidic diet and hyperproteic diet. In whichever case, the common result should be weight loss and reduction of abdominal fat [146].

In recent years, a type of nutritional intervention that has had notable success (thanks to numerous clinical trials) is the "dietary approaches to stop hypertension" (DASH) diet. The DASH diet was formulated for the first time by the National Institute of Health (NIH) in the 1990s, and was object of many research studies [147]. It promoted the ingestion of vegetable proteins, fibers, fresh vegetables, fruits, extra-virgin olive oil and dried fruits, while suggesting a reduction in animal fats, simple sugars and processed meat. Trials have demonstrated how reduced salt consumption potentiated beneficial effects linked to DASH diet, inducing a reduction in systemic arterial pressure in all patients.

Progressive reduction in energy expenditure associated to increased caloric intake translates into weight gain, which finally amounts to obesity [120].

For this reason lifestyle, and specifically physical activity, has a role of primary importance in the maintenance of a healthy status both in primary and secondary prevention [148]. Numerous studies suggest that physical activity has a beneficial effect in subjects affected by AH, to the point of being compared to pharmaceutical intervention [149, 150]. Thanks to recent technologies, it has been possible to develop network meta-analysis (NMA) models able to compare the efficacy of physical activity and pharmacotherapy alone.

A meta-analysis by Naci and Ioannidis [150] has highlighted how physical activity alone elicits similar results to pharmacological therapy in terms of reduction of mortality in hypertensive patients. Such reduction has been studied in patients with coronary heart disease, post-infarct rehabilitation, cardiac insufficiency and in the prevention of diabetes. A recent study by Dempsey et al. [151] has compared seven continuative hours of inactivity to 3 min of light physical activity every 30 min (6 min of physical activity per hour). Such light physical activity, significantly reduced systolic and diastolic arterial blood pressure values, and reduced hematic norepinephrine values.

Thus, it is evident that physical activity plays a pivotal role in the insurgence and management of AH. The challenge for the future will be to identify, using NMAs, different types of physical activity and pharmaceuticals that can be administered in a personalized manner based on the subject's unique characteristics.

Substantial medical literature has tried to identify the mechanisms relating body weight control and cigarette smoke [152].

Even if, an increase in body weight was highlighted following smoking cessation (due to an increase in caloric intake because of the lack of smoking), many studies have also stated that this is a transitory condition [153]. In fact, it has been demonstrated that the weight gain straight after smoking cessation, normalizes in about 6 months with the re-establishment of the normal energetic intake [154].

Alcohol consumption has been a part of food culture since antiquity. Even if there are potential beneficial effects that reside in compounds present in alcoholic beverages, like in red wine, red wine, they have a caloric intake of 7.1 kcal/g of alcohol, therefore not recommended in obese subjects [155]. For this reason, alcohol intake should be controlled and modest, and in order to attain its beneficial potential the quality should also be considered [156].

5. Conclusions

Obesity-related hypertension represents a public health problem, especially as hypertensive-obese subjects has notably enhanced and precocious CV-related morbidity and mortality compared to the general population. It is therefore useful to act as promptly as possible, by intervening with strategies to contrast the insurgence excessive weight gain, obesity and their relative comorbidities such as obesity-related hypertension. It would be optimal to carry out an educational scheme aim at adolescents, in order to educate the population to undertake a correct lifestyle, which will contrast the insurgence of problems cited above later in life. A correct lifestyle is characterized by the combination of constant levels of physical activity and a balanced diet, taking as models the DASH Diet and/or the Mediterranean one. Once a subject develops obesity and hypertension, a pharmaceutical approach (to control systemic arterial blood pressure and obesity) should be flanked to the dietary intervention.

Acknowledgements

We are grateful to Dr Georgia Wilson Jones for the language revision of the manuscript. We are indebted to Dr. Giulia Marrone and Dr Manuela Di Lauro for their technical assistance.

Conflict of interest

The authors declare no conflict of interest.

Author details

Annalisa Noce* and Nicola Di Daniele
Division of Medical Sciences, Internal Medicine-Center of Hypertension and Nephrology Unit, Department of Systems Medicine, Tor Vergata University, Rome, Italy

*Address all correspondence to: annalisa.noce@uniroma2.it

References

[1] World Health Organization. Obesity: Preventing and managing the global epidemic. Report of WHO Consultation. WHO Technical Report Series 894. 2000

[2] Ng M, Fleming T, Robinson M, Thomson B, Graetz N, Margono C, et al. Global, regional, and national prevalence of overweight and obesity in children and adults during 1980-2013: A systematic analysis for the Global Burden of Disease Study 2013. Lancet. 2014;**384**:766-781. DOI: 10.1016/S0140-6736(14)60460-8

[3] Benjamin EJ, Blaha MJ, Chiuve SE, Cushman M, Das SR, Deo R, et al. Heart disease and stroke Statistics-2017 update: A report from the American Heart Association. Circulation. 2017;**135**:e146-e603. DOI: 10.1161/CIR.0000000000000485

[4] DiBonaventura M, Nicolucci A, Meincke H, Le Lay A, Fournier J. Obesity in Germany and Italy: Prevalence, comorbidities, and associations with patient outcomes. ClinicoEconomics & Outcomes Research. 2018;**10**:457-475. DOI: 10.2147/CEOR.S157673

[5] Chooi YC, Ding C, Magkos F. The epidemiology of obesity. Metabolism. 2019;**92**:6-10. DOI: 10.1016/j.metabol.2018.09.005

[6] Fox CS, Pencina MJ, Wilson PW, Paynter NP, Vasan RS, D'Agostino RB Sr. Lifetime risk of cardiovascular disease among individuals with and without diabetes stratified by obesity status in the Framingham heart study. Diabetes Care. 2008;**31**:1582-1584. DOI: 10.2337/dc08-0025

[7] Canale MP, Manca di Villahermosa S, Martino G, Rovella V, Noce A, De Lorenzo A, et al. Obesity-related metabolic syndrome: Mechanisms of sympathetic overactivity. International Journal of Endocrinology. 2013;**2013**:865965. DOI: 10.1155/2013/865965

[8] Landsberg L, Aronne LJ, Beilin LJ, Burke V, Igel LI, Lloyd-Jones D, et al. Obesity-related hypertension: Pathogenesis, cardiovascular risk, and treatment: A position paper of The Obesity Society and the American Society of Hypertension. Journal of Clinical Hypertension (Greenwich, Conn.). 2013;**15**:14-33. DOI: 10.1111/jch.12049

[9] Go AS, Mozaffarian D, Roger VL, Benjamin EJ, Berry JD, Blaha MJ, et al. Executive summary: Heart disease and stroke statistics. 2014 update: A report from the American Heart Association. Circulation. 2014;**129**(3):399-410

[10] Leggio M, Lombardi M, Caldarone E, Severi P, D'Emidio S, Armeni M, et al. The relationship between obesity and hypertension: An updated comprehensive overview on vicious twins. Hypertension Research. 2017;**40**:947-963. DOI: 10.1038/hr.2017.75

[11] Kannel WB, Brand N, Skinner JJ Jr, Dawber TR, McNamara PM. The relation of adiposity to blood pressure and development of hypertension. The Framingham study. Annals of Internal Medicine. 1967;**67**:48-59

[12] Lapidus L, Bengtsson C, Larsson B, Pennert K, Rybo E, Sjostrom L. Distribution of adipose tissue and risk of cardiovascular disease and death: A 12 year follow up of participants in the population study of women in Gothenburg, Sweden. British Medical Journal (Clinical Research Ed.). 1984;**289**:1257-1261

[13] Larsson B, Svardsudd K, Welin L, Wilhelmsen L, Bjorntorp P, Tibblin G.

Abdominal adipose tissue distribution, obesity, and risk of cardiovascular disease and death: 13 year follow up of participants in the study of men born in 1913. British Medical Journal (Clinical Research Ed.). 1984;**288**:1401-1404

[14] Cassano PA, Segal MR, Vokonas PS, Weiss ST. Body fat distribution, blood pressure, and hypertension. A prospective cohort study of men in the normative aging study. Annals of Epidemiology. 1990;**1**:33-48

[15] Vague J. The degree of masculine differentiation of obesities: A factor determining predisposition to diabetes, atherosclerosis, gout, and uric calculous disease. American Journal of Clinical Nutrition. 1956;**4**:20-34. DOI: 10.1093/ajcn/4.1.20

[16] Kissebah AH, Vydelingum N, Murray R, Evans DJ, Hartz AJ, Kalkhoff RK, et al. Relation of body fat distribution to metabolic complications of obesity. The Journal of Clinical Endocrinology and Metabolism. 1982;**54**:254-260. DOI: 10.1210/jcem-54-2-254

[17] Krotkiewski M, Bjorntorp P, Sjostrom L, Smith U. Impact of obesity on metabolism in men and women. Importance of regional adipose tissue distribution. The Journal of Clinical Investigation. 1983;**72**:1150-1162. DOI: 10.1172/JCI111040

[18] Ferrannini E, Buzzigoli G, Bonadonna R, Giorico MA, Oleggini M, Graziadei L, et al. Insulin resistance in essential hypertension. The New England Journal of Medicine. 1987;**317**:350-357. DOI: 10.1056/NEJM198708063170605

[19] Modan M, Halkin H, Almog S, Lusky A, Eshkol A, Shefi M, et al. Hyperinsulinemia. A link between hypertension obesity and glucose intolerance. The Journal of Clinical Investigation. 1985;**75**:809-817. DOI: 10.1172/JCI111776

[20] Wirix AJ, Kaspers PJ, Nauta J, Chinapaw MJ, Kist-van Holthe JE. Pathophysiology of hypertension in obese children: A systematic review. Obesity Reviews. 2015;**16**:831-842. DOI: 10.1111/obr.12305

[21] Hall JE, do Carmo JM, da Silva AA, Wang Z, Hall ME. Obesity-induced hypertension: Interaction of neurohumoral and renal mechanisms. Circulation Research. 2015;**116**:991-1006. DOI: 10.1161/CIRCRESAHA.116.305697

[22] Kotsis V, Stabouli S, Papakatsika S, Rizos Z, Parati G. Mechanisms of obesity-induced hypertension. Hypertension Research. 2010;**33**: 386-393. DOI: 10.1038/hr.2010.9

[23] Becton LJ, Shatat IF, Flynn JT. Hypertension and obesity: Epidemiology, mechanisms and clinical approach. Indian Journal of Pediatrics. 2012;**79**:1056-1061. DOI: 10.1007/s12098-012-0777-x

[24] Flynn J. The changing face of pediatric hypertension in the era of the childhood obesity epidemic. Pediatric Nephrology. 2013;**28**:1059-1066. DOI: 10.1007/s00467-012-2344-0

[25] Grassi G, Seravalle G, Dell'Oro R, Turri C, Bolla GB, Mancia G. Adrenergic and reflex abnormalities in obesity-related hypertension. Hypertension. 2000;**36**:538-542

[26] Grassi G, Dell'Oro R, Facchini A, Quarti Trevano F, Bolla GB, Mancia G. Effect of central and peripheral body fat distribution on sympathetic and baroreflex function in obese normotensives. Journal of Hypertension. 2004;**22**:2363-2369

[27] Alvarez GE, Ballard TP, Beske SD, Davy KP. Subcutaneous obesity is not associated with sympathetic neural activation. American Journal of Physiology. Heart and Circulatory

Physiology. 2004;**287**:H414-H418. DOI: 10.1152/ajpheart.01046.2003

[28] Canoy D, Luben R, Welch A, Bingham S, Wareham N, Day N, et al. Fat distribution, body mass index and blood pressure in 22,090 men and women in the Norfolk cohort of the European Prospective Investigation into Cancer and Nutrition (EPIC-Norfolk) study. Journal of Hypertension. 2004;**22**:2067-2074

[29] Hall JE, da Silva AA, do Carmo JM, Dubinion J, Hamza S, Munusamy S, et al. Obesity-induced hypertension: Role of sympathetic nervous system, leptin, and melanocortins. The Journal of Biological Chemistry. 2010;**285**:17271-17276. DOI: 10.1074/jbc. R110.113175

[30] Lohmeier TE, Iliescu R. The sympathetic nervous system in obesity hypertension. Current Hypertension Reports. 2013;**15**:409-416. DOI: 10.1007/ s11906-013-0356-1

[31] Aghamohammadzadeh R, Heagerty AM. Obesity-related hypertension: Epidemiology, pathophysiology, treatments, and the contribution of perivascular adipose tissue. Annals of Medicine. 2012;**44**(Suppl 1):S74-S84. DOI: 10.3109/07853890.2012.663928

[32] DeMarco VG, Aroor AR, Sowers JR. The pathophysiology of hypertension in patients with obesity. Nature Reviews. Endocrinology. 2014;**10**:364-376. DOI: 10.1038/ nrendo.2014.44

[33] Rumantir MS, Vaz M, Jennings GL, Collier G, Kaye DM, Seals DR, et al. Neural mechanisms in human obesity-related hypertension. Journal of Hypertension. 1999;**17**:1125-1133

[34] Kassab S, Kato T, Wilkins FC, Chen R, Hall JE, Granger JP. Renal denervation attenuates the sodium retention and hypertension associated with obesity. Hypertension. 1995;**25**:893-897

[35] Lansberg L. Hypertension in the Twentieth Century: Concepts and Achievements: Handbook of Hypertension. Vol. 22. Amsterdam, Netherlands: Elsevier; 2004. pp. 245-261

[36] Hausberg M, Mark AL, Hoffman RP, Sinkey CA, Anderson EA. Dissociation of sympathoexcitatory and vasodilator actions of modestly elevated plasma insulin levels. Journal of Hypertension. 1995;**13**:1015-1021

[37] Grassi G, Seravalle G, Colombo M, Bolla G, Cattaneo BM, Cavagnini F, et al. Body weight reduction, sympathetic nerve traffic, and arterial baroreflex in obese normotensive humans. Circulation. 1998;**97**:2037-2042

[38] DeFronzo RA. Insulin and renal sodium handling: Clinical implications. International Journal of Obesity. 1981;**5**(suppl 1):93-104

[39] Kennedy A, Gettys TW, Watson P, Wallace P, Ganaway E, Pan Q, et al. The metabolic significance of leptin in humans: Gender-based differences in relationship to adiposity, insulin sensitivity, and energy expenditure. The Journal of Clinical Endocrinology and Metabolism. 1997;**82**:1293-1300. DOI: 10.1210/jcem.82.4.3859

[40] Tang-Christensen M, Havel PJ, Jacobs RR, Larsen PJ, Cameron JL. Central administration of leptin inhibits food intake and activates the sympathetic nervous system in rhesus macaques. The Journal of Clinical Endocrinology and Metabolism. 1999;**84**:711-717. DOI: 10.1210/ jcem.84.2.5458

[41] Azar ST, Salti I, Zantout MS, Shahine CH, Zalloua PA. Higher serum leptin level in women than in men

with type 1 diabetes. The American Journal of the Medical Sciences. 2002;**323**:206-209

[42] Chrysant SG. Pathophysiology and treatment of obesity-related hypertension. Journal of Clinical Hypertension (Greenwich). 2019;**21**(5):555-559. DOI: 10.1111/jch.13518

[43] Shek EW, Brands MW, Hall JE. Chronic leptin infusion increases arterial pressure. Hypertension. 1998;**31**:409-414

[44] Dunbar JC, Hu Y, Lu H. Intracerebroventricular leptin increases lumbar and renal sympathetic nerve activity and blood pressure in normal rats. Diabetes. 1997;**46**:2040-2043

[45] Landsberg L. Diet, obesity and hypertension: An hypothesis involving insulin, the sympathetic nervous system, and adaptive thermogenesis. The Quarterly Journal of Medicine. 1986;**61**:1081-1090

[46] Bentley-Lewis R, Adler GK, Perlstein T, Seely EW, Hopkins PN, Williams GH, et al. Body mass index predicts aldosterone production in normotensive adults on a high-salt diet. The Journal of Clinical Endocrinology and Metabolism. 2007;**92**:4472-4475. DOI: 10.1210/jc.2007-1088

[47] Goodfriend TL, Kelley DE, Goodpaster BH, Winters SJ. Visceral obesity and insulin resistance are associated with plasma aldosterone levels in women. Obesity Research. 1999;**7**:355-362

[48] Nagase M, Yoshida S, Shibata S, Nagase T, Gotoda T, Ando K, et al. Enhanced aldosterone signaling in the early nephropathy of rats with metabolic syndrome: Possible contribution of fat-derived factors. Journal of the American Society of Nephrology. 2006;**17**:3438-3446. DOI: 10.1681/ASN.2006080944

[49] Jeon JH, Kim KY, Kim JH, Baek A, Cho H, Lee YH, et al. A novel adipokine CTRP1 stimulates aldosterone production. The FASEB Journal. 2008;**22**:1502-1511. DOI: 10.1096/fj.07-9412com

[50] Kawarazaki W, Fujita T. The role of aldosterone in obesity-related hypertension. American Journal of Hypertension. 2016;**29**:415-423. DOI: 10.1093/ajh/hpw003

[51] Engeli S, Sharma AM. The renin-angiotensin system and natriuretic peptides in obesity-associated hypertension. Journal of Molecular Medicine (Berlin, Germany). 2001;**79**:21-29

[52] Marcus Y, Shefer G, Stern N. Adipose tissue renin-angiotensin-aldosterone system (RAAS) and progression of insulin resistance. Molecular and Cellular Endocrinology. 2013;**378**:1-14. DOI: 10.1016/j.mce.2012.06.021

[53] Cassis LA, Police SB, Yiannikouris F, Thatcher SE. Local adipose tissue renin-angiotensin system. Current Hypertension Reports. 2008;**10**:93-98

[54] Yasue S, Masuzaki H, Okada S, Ishii T, Kozuka C, Tanaka T, et al. Adipose tissue-specific regulation of angiotensinogen in obese humans and mice: Impact of nutritional status and adipocyte hypertrophy. American Journal of Hypertension. 2010;**23**:425-431. DOI: 10.1038/ajh.2009.263

[55] Yiannikouris F, Gupte M, Putnam K, Thatcher S, Charnigo R, Rateri DL, et al. Adipocyte deficiency of angiotensinogen prevents obesity-induced hypertension in male mice. Hypertension. 2012;**60**:1524-1530. DOI: 10.1161/HYPERTENSIONAHA.112.192690

[56] Karlsson C, Lindell K, Ottosson M, Sjostrom L, Carlsson B,

Carlsson LM. Human adipose tissue expresses angiotensinogen and enzymes required for its conversion to angiotensin II. The Journal of Clinical Endocrinology and Metabolism. 1998;**83**:3925-3929. DOI: 10.1210/jcem.83.11.5276

[57] Seravalle G, Colombo M, Perego P, Giardini V, Volpe M, Dell'Oro R, et al. Long-term sympathoinhibitory effects of surgically induced weight loss in severe obese patients. Hypertension. 2014;**64**:431-437. DOI: 10.1161/HYPERTENSIONAHA.113.02988

[58] Tuck ML, Sowers J, Dornfeld L, Kledzik G, Maxwell M. The effect of weight reduction on blood pressure, plasma renin activity, and plasma aldosterone levels in obese patients. The New England Journal of Medicine. 1981;**304**:930-933. DOI: 10.1056/NEJM198104163041602

[59] Rocchini AP, Katch VL, Grekin R, Moorehead C, Anderson J. Role for aldosterone in blood pressure regulation of obese adolescents. The American Journal of Cardiology. 1986;**57**:613-618

[60] Engeli S, Bohnke J, Gorzelniak K, Janke J, Schling P, Bader M, et al. Weight loss and the renin-angiotensin-aldosterone system. Hypertension. 2005;**45**:356-362. DOI: 10.1161/01.HYP.0000154361.47683.d3

[61] Ruano M, Silvestre V, Castro R, Garcia-Lescun MC, Rodriguez A, Marco A, et al. Morbid obesity, hypertensive disease and the renin-angiotensin-aldosterone axis. Obesity Surgery. 2005;**15**:670-676. DOI: 10.1381/0960892053923734

[62] Goodfriend TL, Ball DL, Egan BM, Campbell WB, Nithipatikom K. Epoxy-keto derivative of linoleic acid stimulates aldosterone secretion. Hypertension. 2004;**43**:358-363. DOI: 10.1161/01.HYP.0000113294.06704.64

[63] Naguib MT. Kidney disease in the obese patient. Southern Medical Journal. 2014;**107**:481-485. DOI: 10.14423/SMJ.0000000000000141

[64] Hall JE, Brands MW, Dixon WN, Smith MJ Jr. Obesity-induced hypertension. Renal function and systemic hemodynamics. Hypertension. 1993;**22**:292-299

[65] Tchernof A, Despres JP. Pathophysiology of human visceral obesity: An update. Physiological Reviews. 2013;**93**:359-404. DOI: 10.1152/physrev.00033.2011

[66] Carroll JF, Huang M, Hester RL, Cockrell K, Mizelle HL. Hemodynamic alterations in hypertensive obese rabbits. Hypertension. 1995;**26**:465-470

[67] Hall ME, do Carmo JM, da Silva AA, Juncos LA, Wang Z, Hall JE. Obesity, hypertension, and chronic kidney disease. International Journal of Nephrology and Renovascular Disease. 2014;**7**:75-88. DOI: 10.2147/IJNRD.S39739

[68] Ahmed SB, Fisher ND, Stevanovic R, Hollenberg NK. Body mass index and angiotensin-dependent control of the renal circulation in healthy humans. Hypertension. 2005;**46**:1316-1320. DOI: 10.1161/01.HYP.0000190819.07663.da

[69] Frigolet ME, Torres N, Tovar AR. The renin-angiotensin system in adipose tissue and its metabolic consequences during obesity. The Journal of Nutritional Biochemistry. 2013;**24**:2003-2015. DOI: 10.1016/j.jnutbio.2013.07.002

[70] Sun K, Tordjman J, Clement K, Scherer PE. Fibrosis and adipose tissue dysfunction. Cell Metabolism. 2013;**18**:470-477. DOI: 10.1016/j.cmet.2013.06.016

[71] Guzik TJ, Skiba DS, Touyz RM, Harrison DG. The role of infiltrating immune cells in dysfunctional adipose

tissue. Cardiovascular Research. 2017;**113**:1009-1023. DOI: 10.1093/cvr/cvx108

[72] Ronti T, Lupattelli G, Mannarino E. he endocrine function of adipose tissue: An update. Clinical Endocrinology. 2006;**64**:355-365. DOI: 10.1111/j.1365-2265.2006.02474.x

[73] Mikolajczyk TP, Nosalski R, Szczepaniak P, Budzyn K, Osmenda G, Skiba D, et al. Role of chemokine RANTES in the regulation of perivascular inflammation, T-cell accumulation, and vascular dysfunction in hypertension. The FASEB Journal. 2016;**30**:1987-1999. DOI: 10.1096/fj.201500088R

[74] Marvar PJ, Thabet SR, Guzik TJ, Lob HE, McCann LA, Weyand C, et al. Central and peripheral mechanisms of T-lymphocyte activation and vascular inflammation produced by angiotensin II-induced hypertension. Circulation Research. 2010;**107**:263-270. DOI: 10.1161/CIRCRESAHA.110.217299

[75] Wilk G, Osmenda G, Matusik P, Nowakowski D, Jasiewicz-Honkisz B, Ignacak A, et al. Endothelial function assessment in atherosclerosis: Comparison of brachial artery flowmediated vasodilation and peripheral arterial tonometry. Polskie Archiwum Medycyny Wewnętrznej. 2013;**123**:443-452

[76] Guzik TJ, Hoch NE, Brown KA, McCann LA, Rahman A, Dikalov S, et al. Role of the T cell in the genesis of angiotensin II induced hypertension and vascular dysfunction. The Journal of Experimental Medicine. 2007;**204**:2449-2460. DOI: 10.1084/jem.20070657

[77] Nosalski R, McGinnigle E, Siedlinski M, Guzik TJ. Novel immune mechanisms in hypertension and cardiovascular risk. Current Cardiovascular Risk Reports. 2017;**11**:12. DOI: 10.1007/s12170-017-0537-6

[78] Guzik TJ, Olszanecki R, Sadowski J, Kapelak B, Rudzinski P, Jopek A, et al. Superoxide dismutase activity and expression in human venous and arterial bypass graft vessels. Journal of Physiology and Pharmacology. 2005;**56**:313-323

[79] Coccagna G, Mantovani M, Brignani F, Parchi C, Lugaresi E. Continuous recording of the pulmonary and systemic arterial pressure during sleep in syndromes of hypersomnia with periodic breathing. Bulletin de Physio-Pathologie Respiratoire. 1972;**8**:1159-1172

[80] Marcus JA, Pothineni A, Marcus CZ, Bisognano JD. The role of obesity and obstructive sleep apnea in the pathogenesis and treatment of resistant hypertension. Current Hypertension Reports. 2014;**16**:411. DOI: 10.1007/s11906-013-0411-y

[81] Flegal KM, Carroll MD, Ogden CL, Curtin LR. Prevalence and trends in obesity among US adults, 1999-2008. JAMA. 2010;**303**:235-241. DOI: 10.1001/jama.2009.2014

[82] Parati G, Pengo MF, Lombardi C. Obstructive sleep apnea and hypertension: Why treatment does not consistently improve blood pressure. Current Hypertension Reports. 2019;**21**:30. DOI: 10.1007/s11906-019-0935-x

[83] Iturriaga R, Oyarce MP, Dias ACR. Role of carotid body in intermittent hypoxia-related hypertension. Current Hypertension Reports. 2017;**19**:38. DOI: 10.1007/s11906-017-0735-0

[84] Khan A, Patel NK, O'Hearn DJ, Khan S. Resistant hypertension and obstructive sleep apnea. International Journal of Hypertension. 2013;**2013**:193010. DOI: 10.1155/2013/193010

[85] Phillips BG, Narkiewicz K, Pesek CA, Haynes WG, Dyken ME, Somers VK.

Effects of obstructive sleep apnea on endothelin-1 and blood pressure. Journal of Hypertension. 1999;**17**:61-66

[86] Belaidi E, Joyeux-Faure M, Ribuot C, Launois SH, Levy P, Godin-Ribuot D. Major role for hypoxia inducible factor-1 and the endothelin system in promoting myocardial infarction and hypertension in an animal model of obstructive sleep apnea. Journal of the American College of Cardiology. 2009;**53**:1309-1317. DOI: 10.1016/j.jacc.2008.12.050

[87] Hla KM, Young T, Finn L, Peppard PE, Szklo-Coxe M, Stubbs M. Longitudinal association of sleep-disordered breathing and nondipping of nocturnal blood pressure in the Wisconsin Sleep Cohort Study. Sleep. 2008;**31**:795-800

[88] James PA, Oparil S, Carter BL, Cushman WC, Dennison-Himmelfarb C, Handler J, et al. 2014 evidence-based guideline for the management of high blood pressure in adults: Report from the panel members appointed to the Eighth Joint National Committee (JNC 8). JAMA. 2014;**311**:507-520. DOI: 10.1001/jama.2013.284427

[89] Williams B, Mancia G, Spiering W, Agabiti Rosei E, Azizi M, Burnier M, et al. 2018 ESC/ESH guidelines for the management of arterial hypertension. European Heart Journal. 2018;**39**: 3021-3104. DOI: 10.1093/eurheartj/ehy339

[90] Iqbal AM, Jamal SF. Essential Hypertension. Treasure Island (FL): StatPearls; 2019

[91] Jandeleit-Dahm KA, Tikellis C, Reid CM, Johnston CI, Cooper ME. Why blockade of the renin-angiotensin system reduces the incidence of new-onset diabetes. Journal of Hypertension. 2005;**23**:463-473

[92] Sharma AM, Engeli S. The role of renin-angiotensin system

blockade in the management of hypertension associated with the cardiometabolic syndrome. Journal of the Cardiometabolic Syndrome. 2006;**1**:29-35

[93] Gupta AK, Dahlof B, Dobson J, Sever PS, Wedel H, Poulter NR. Anglo-Scandinavian Cardiac Outcomes Trial, I. Determinants of new-onset diabetes among 19,257 hypertensive patients randomized in the Anglo-Scandinavian Cardiac Outcomes Trial--Blood Pressure Lowering rm and the relative influence of antihypertensive medication. Diabetes Care. 2008;**31**:982-988. DOI: 10.2337/dc07-1768

[94] Lardizabal JA, Deedwania PC. The role of renin-angiotensin agents in altering the natural history of type 2 diabetes mellitus. Current Cardiology Reports. 2010;**12**:464-471. DOI: 10.1007/s11886-010-0138-1

[95] Lavie CJ, Patel DA, Milani RV, Ventura HO, Shah S, Gilliland Y. Impact of echocardiographic left ventricular geometry on clinical prognosis. Progress in Cardiovascular Diseases. 2014;**57**:3-9. DOI: 10.1016/j.pcad.2014.05.003

[96] Artham SM, Lavie CJ, Milani RV, Patel DA, Verma A, Ventura HO. Clinical impact of left ventricular hypertrophy and implications for regression. Progress in Cardiovascular Diseases. 2009;**52**:153-167. DOI: 10.1016/j.pcad.2009.05.002

[97] Wing LM, Reid CM, Ryan P, Beilin LJ, Brown MA, Jennings GL, et al. A comparison of outcomes with angiotensin-converting--enzyme inhibitors and diuretics for hypertension in the elderly. The New England Journal of Medicine. 2003;**348**:583-592. DOI: 10.1056/NEJMoa021716

[98] The Seventh Report of the Joint National Committee on Prevention, Detection, Evaluation, and Treatment

of High Blood Pressure. USA: Bethesda (MD); 2004

[99] Cooper-DeHoff RM, Wen S, Beitelshees AL, Zineh I, Gums JG, Turner ST, et al. Impact of abdominal obesity on incidence of adverse metabolic effects associated with antihypertensive medications. Hypertension. 2010;**55**:61-68. DOI: 10.1161/ HYPERTENSIONAHA.109.139592

[100] Torre JJ, Bloomgarden ZT, Dickey RA, Hogan MJ, Janick JJ, Jyothinagaram SG, et al. American Association of Clinical Endocrinologists Medical Guidelines for Clinical Practice for the diagnosis and treatment of hypertension. Endocrine Practice. 2006;**12**:193-222

[101] Habibi J, Whaley-Connell A, Hayden MR, DeMarco VG, Schneider R, Sowers SD, et al. Renin inhibition attenuates insulin resistance, oxidative stress, and pancreatic remodeling in the transgenic Ren2 rat. Endocrinology. 2008;**149**:5643-5653. DOI: 10.1210/ en.2008-0070

[102] Lee P, Kengne AP, Greenfield JR, Day RO, Chalmers J, Ho KK. Metabolic sequelae of beta-blocker therapy: Weighing in on the obesity epidemic? International Journal of Obesity. 2011;**35**:1395-1403. DOI: 10.1038/ ijo.2010.284

[103] Messerli FH, Bell DS, Fonseca V, Katholi RE, McGill JB, Phillips RA, et al. Body weight changes with beta-blocker use: Results from GEMINI. The American Journal of Medicine. 2007;**120**:610-615. DOI: 10.1016/j. amjmed.2006.10.017

[104] Kidambi S, Kotchen TA. Treatment of hypertension in obese patients. American Journal of Cardiovascular Drugs. 2013;**13**:163-175. DOI: 10.1007/ s40256-013-0008-5

[105] Schmieder RE, Gatzka C, Schachinger H, Schobel H, Ruddel H. Obesity as a determinant for response to antihypertensive treatment. BMJ. 1993;**307**:537-540. DOI: 10.1136/ bmj.307.6903.537

[106] Stoa-Birketvedt G, Thom E, Aarbakke J, Florholmen J. Body fat as a predictor of the antihypertensive effect of nifedipine. Journal of Internal Medicine. 1995;**237**:169-173

[107] Zhang R, Thakur V, Morse S, Reisin E. Renal and cardiovascular considerations for the nonpharmacological and pharmacological therapies of obesity-hypertension. Journal of Human Hypertension. 2002;**16**:819-827. DOI: 10.1038/sj.jhh.1001496

[108] Bakris GL, Copley JB, Vicknair N, Sadler R, Leurgans S. Calcium channel blockers versus other antihypertensive therapies on progression of NIDDM associated nephropathy. Kidney International. 1996;**50**:1641-1650

[109] Hummel D, Raff U, Schwarz TK, Schneider MP, Schmieder RE, Schmidt BM. Dihydropyridine calcium antagonists are associated with increased albuminuria in treatment-resistant hypertensives. Journal of Nephrology. 2010;**23**:563-568

[110] Gress TW, Nieto FJ, Shahar E, Wofford MR, Brancati FL. Hypertension and antihypertensive therapy as risk factors for type 2 diabetes mellitus. Atherosclerosis Risk in Communities Study. The New England Journal of Medicine. 2000;**342**:905-912. DOI: 10.1056/NEJM200003303421301

[111] Weber MA, Jamerson K, Bakris GL, Weir MR, Zappe D, Zhang Y, et al. Effects of body size and hypertension treatments on cardiovascular event rates: Subanalysis of the ACCOMPLISH randomised controlled trial. Lancet.

2013;**381**:537-545. DOI: 10.1016/
S0140-6736(12)61343-9

[112] Dahlof B, Sever PS, Poulter NR, Wedel H, Beevers DG, Caulfield M, et al. Prevention of cardiovascular events with an antihypertensive regimen of amlodipine adding perindopril as required versus atenolol adding bendroflumethiazide as required, in the Anglo-Scandinavian Cardiac Outcomes Trial-Blood Pressure Lowering Arm (ASCOT-BPLA): A multicentre randomised controlled trial. Lancet. 2005;**366**:895-906. DOI: 10.1016/
S0140-6736(05)67185-1

[113] Wofford MR, Smith G, Minor DS. The treatment of hypertension in obese patients. Current Hypertension Reports. 2008;**10**:143-150

[114] Noce A, Marrone G, Rovella V, Busca A, Gola C, Ferrannini M, et al. Fenoldopam mesylate: A narrative review of its use in acute kidney injury. Current Pharmaceutical Biotechnology. 2019;**20**(5). DOI: 10.2174/1389201020666 190417124711

[115] Rovella V, Ferrannini M, Tesauro M, Marrone G, Busca A, Sorge R, et al. Effects of fenoldopam on renal blood flow in hypertensive chronic kidney disease. Journal of Nephrology. 2019;**32**:75-81. DOI: 10.1007/
s40620-018-0496-0

[116] Cataldi M, di Geronimo O, Trio R, Scotti A, Memoli A, Capone D, et al. Utilization of antihypertensive drugs in obesity-related hypertension: A retrospective observational study in a cohort of patients from Southern Italy. BMC Pharmacology and Toxicology. 2016;**17**:9. DOI: 10.1186/
s40360-016-0055-z

[117] NHLBI Obesity Education Initiative Expert Panel on the Identification E, Treatment of Obesity in Adults (US). Clinical Guidelines on the Identification, Evaluation, and Treatment of Overweight and Obesity in Adults. USA: National Heart, Lung, and Blood Institute; 1998

[118] Snow V, Barry P, Fitterman N, Qaseem A, Weiss K. Clinical Efficacy Assessment Subcommittee of the American College of, P. Pharmacologic and surgical management of obesity in primary care: A clinical practice guideline from the American College of Physicians. Annals of Internal Medicine. 2005;**142**:525-531

[119] Scheen AJ. Sibutramine on cardiovascular outcome. Diabetes Care. 2011;**34**(Suppl 2):S114-S119. DOI: 10.2337/dc11-s205

[120] Derosa G, Cicero AF, Murdolo G, Piccinni MN, Fogari E, Bertone G, et al. Efficacy and safety comparative evaluation of orlistat and sibutramine treatment in hypertensive obese patients. Diabetes, Obesity & Metabolism. 2005;**7**:47-55. DOI: 10.1111/j.1463-1326.2004.00372.x

[121] Over-the-counter weight loss with orlistat? Evidence-Based Nursing. 2010;**13**:98-100. DOI: 10.1136/
dtb.2009.10.0046

[122] Guerciolini R. Mode of action of orlistat. International Journal of Obesity and Related Metabolic Disorders. 1997;**21**(Suppl 3):S12-S23

[123] Rossner S, Sjostrom L, Noack R, Meinders AE, Noseda G. Weight loss, weight maintenance, and improved cardiovascular risk factors after 2 years treatment with orlistat for obesity. European Orlistat Obesity Study Group. Obesity Research. 2000;**8**:49-61. DOI: 10.1038/oby.2000.8

[124] Brown SA, Upchurch S, Anding R, Winter M, Ramirez G. Promoting weight loss in type II diabetes. Diabetes Care. 1996;**19**:613-624

[125] Hauner H, Petzinna D, Sommerauer B, Toplak H. Effect of acarbose on weight maintenance after dietary weight loss in obese subjects. Diabetes, Obesity & Metabolism. 2001;**3**:423-427

[126] Lawrence CB, Turnbull AV, Rothwell NJ. Hypothalamic control of feeding. Current Opinion in Neurobiology. 1999;**9**:778-783

[127] Garfield AS, Lam DD, Marston OJ, Przydzial MJ, Heisler LK. Role of central melanocortin pathways in energy homeostasis. Trends in Endocrinology and Metabolism. 2009;**20**:203-215. DOI: 10.1016/j.tem.2009.02.002

[128] Smitka K, Papezova H, Vondra K, Hill M, Hainer V, Nedvidkova J. The role of "mixed" orexigenic and anorexigenic signals and autoantibodies reacting with appetite-regulating neuropeptides and peptides of the adipose tissue-gut-brain axis: Relevance to food intake and nutritional status in patients with anorexia nervosa and bulimia nervosa. International Journal of Endocrinology. 2013;**2013**:483145. DOI: 10.1155/2013/483145

[129] Jollis JG, Landolfo CK, Kisslo J, Constantine GD, Davis KD, Ryan T. Fenfluramine and phentermine and cardiovascular findings: Effect of treatment duration on prevalence of valve abnormalities. Circulation. 2000;**101**:2071-2077

[130] Gustafson A, King C, Rey JA. Lorcaserin (Belviq): A selective serotonin 5-HT2C agonist In the treatment of obesity. PT. 2013;**38**:525-534

[131] Smith SR, Weissman NJ, Anderson CM, Sanchez M, Chuang E, Stubbe S, et al. Multicenter, placebo-controlled trial of lorcaserin for weight management. The New England Journal of Medicine. 2010;**363**:245-256. DOI: 10.1056/NEJMoa0909809

[132] Ioannides-Demos LL, Piccenna L, McNeil JJ. Pharmacotherapies for obesity: Past, current, and future therapies. Journal of Obesity. 2011;**2011**:179674. DOI: 10.1155/2011/179674

[133] Kaplan LM. Pharmacological therapies for obesity. Gastroenterology Clinics of North America. 2005;**34**: 91-104. DOI: 10.1016/j.gtc.2004.12.002

[134] Patel DK, Stanford FC. Safety and tolerability of new-generation anti-obesity medications: A narrative review. Postgraduate Medicine. 2018;**130**:173-182. DOI: 10.1080/00325481.2018.1435129

[135] Levri KM, Slaymaker E, Last A, Yeh J, Ference J, D'Amico F, et al. Metformin as treatment for overweight and obese adults: A systematic review. Annals of Family Medicine. 2005;**3**: 457-461. DOI: 10.1370/afm.343

[136] Diabetes Prevention Program Research, G. The diabetes prevention program (DPP): Description of lifestyle intervention. Diabetes Care. 2002;**25**:2165-2171

[137] Robinson AC, Burke J, Robinson S, Johnston DG, Elkeles RS. The effects of metformin on glycemic control and serum lipids in insulin-treated NIDDM patients with suboptimal metabolic control. Diabetes Care. 1998;**21**:701-705

[138] Kim W, Egan JM. The role of incretins in glucose homeostasis and diabetes treatment. Pharmacological Reviews. 2008;**60**:470-512. DOI: 10.1124/pr.108.000604

[139] Brown NJ. Cardiovascular effects of antidiabetic agents: Focus on blood pressure effects of incretin-based therapies. Journal of the American Society of Hypertension. 2012;**6**: 163-168. DOI: 10.1016/j. jash.2012.02.003

[140] Horton ES, Silberman C, Davis KL, Berria R. Weight loss, glycemic control, and changes in cardiovascular biomarkers in patients with type 2 diabetes receiving incretin therapies or insulin in a large cohort database. Diabetes Care. 2010;**33**:1759-1765. DOI: 10.2337/dc09-2062

[141] Jiang SZ, Lu W, Zong XF, Ruan HY, Liu Y. Obesity and hypertension. Experimental and Therapeutic Medicine. 2016;**12**:2395-2399. DOI: 10.3892/etm.2016.3667

[142] Collaboration, N.C.D.R.F. Worldwide trends in body-mass index, underweight, overweight, and obesity from 1975 to 2016: A pooled analysis of 2416 population-based measurement studies in 128.9 million children, adolescents, and adults. Lancet. 2017;**390**(17):2627, 32129-2642, 32123. DOI: 10.1016/S0140-6736

[143] Kotchen TA. Obesity-related hypertension: Epidemiology, pathophysiology, and clinical management. American Journal of Hypertension. 2010;**23**:1170-1178. DOI: 10.1038/ajh.2010.172

[144] Racette SB, Deusinger SS, Strube MJ, Highstein GR, Deusinger RH. Weight changes, exercise, and dietary patterns during freshman and sophomore years of college. Journal of American College Health. 2005;**53**:245-251. DOI: 10.3200/JACH.53.6.245-251

[145] Mirowsky J, Ross CE. Social Causes of Psychological Distress. 2nd edition. New York, USA: Aldine Transaction; 2003

[146] Bray GA. Lifestyle and pharmacological approaches to weight loss: Efficacy and safety. The Journal of Clinical Endocrinology and Metabolism. 2008;**93**:S81-S88. DOI: 10.1210/jc.2008-1294

[147] Your Guide to Lowering Your Blood Pressure with DASH. U.S. Department of Health and Human Services National Institutes of Health National Heart, Lung, and Blood Institute. Bethesda, MD, USA: NHLBI Publications and Resources; 2006

[148] Ding D, Lawson KD, Kolbe-Alexander TL, Finkelstein EA, Katzmarzyk PT, van Mechelen W, et al. The economic burden of physical inactivity: A global analysis of major non-communicable diseases. Lancet. 2016;**388**:1311-1324. DOI: 10.1016/S0140-6736(16)30383-X

[149] Cornelissen VA, Smart NA. Exercise training for blood pressure: A systematic review and meta-analysis. Journal of the American Heart Association. 2013;**2**:e004473. DOI: 10.1161/JAHA.112.004473

[150] Naci H, Ioannidis JP. Comparative effectiveness of exercise and drug interventions on mortality outcomes: Metaepidemiological study. British Journal of Sports Medicine. 2015;**49**:1414-1422. DOI: 10.1136/bjsports-2015-f5577rep

[151] Dempsey PC, Sacre JW, Larsen RN, Straznicky NE, Sethi P, Cohen ND, et al. Interrupting prolonged sitting with brief bouts of light walking or simple resistance activities reduces resting blood pressure and plasma noradrenaline in type 2 diabetes. Journal of Hypertension. 2016;**34**:2376-2382. DOI: 10.1097/HJH.0000000000001101

[152] The Health Benefits of Smoking Cessation: A Report of the Surgeon General. United States: Public Health Service. Rockvllle, Maryland USA: Office of the Surgeon General; 1990

[153] Bush T, Lovejoy JC, Deprey M, Carpenter KM. The effect of tobacco cessation on weight gain, obesity, and diabetes risk. Obesity (Silver Spring). 2016;**24**:1834-1841. DOI: 10.1002/oby.21582

[154] Gruber J, Frakes M. Does falling smoking lead to rising obesity? Journal of Health Economics. 2006;**25**:183-197; discussion 389-193. DOI: 10.1016/j. jhealeco.2005.07.005

[155] Traversy G, Chaput JP. Alcohol consumption and obesity: An update. Current Obesity Reports. 2015;**4**:122-130. DOI: 10.1007/s13679-014-0129-4

[156] Bendsen NT, Christensen R, Bartels EM, Kok FJ, Sierksma A, Raben A, et al. Is beer consumption related to measures of abdominal and general obesity? A systematic review and meta-analysis. Nutrition Reviews. 2013;**71**:67-87. DOI: 10.1111/j.1753-4887.2012.00548.x

Chapter 5

Predictors of Resistance Hypertension and Achievement of Target Blood Pressure Levels in Patients with Resistant Hypertension

Yuriy Mykolayovych Sirenko, Oksana Leonidivna Rekovets and Olena Oleksandrivna Torbas

Abstract

Uncontrolled arterial pressure is associated with a fourfold increase in the risk of developing cardiovascular events compared to patients with hypertension who have reached the target blood pressure level. The aim of this study is to evaluate the characteristics of patients with resistant arterial hypertension undergoing inpatient treatment at the Department of Symptomatic Hypertension and assess the prevalence of true resistant hypertension in a cohort of patients who take 3 and more antihypertensive agents, the clinical predictors of resistant hypertension. The study included 1146 patients with resistant AH who received 3 or more antihypertensive drugs with the level of office blood pressure at admission ≥140/90 mm Hg. Patients were followed by the next examinations: body height and body measurements, office blood pressure, echocardiography, sleep apnea determination, blood biochemical analysis, determination of levels of TTH, T3, T4, blood renin, blood aldosterone, metanephrine urine, and cortisol. Our data showed that 31% of patients who received 3 or more antihypertensive drugs had true resistant hypertension. Fixed combinations were taken by 71.9% of patients. We have found which factors were significantly associated with the treatment regimen with ≥3 or 4 drugs. Also we have demonstrated predictors for blood pressure reduction.

Keywords: resistant hypertension, pharmacology, predictors of resistance hypertension, target blood pressure, anti-hypertensive drugs

1. Background

The prevalence of resistant hypertension is very different according to different studies. In an analysis of National Health and Nutrition Examination Survey (NHANES) participants being treated for hypertension, only 53% were controlled to 140/90 mm Hg [1]. In Framingham Heart Study participants, only 48% of treated patients were controlled to 140/90 mm Hg, and less than 40% of elderly participants (75 years of age) were at a goal blood pressure [2]. Among higher-risk populations and, in particular, with application of the lower goal blood pressures recommended in the Seventh Report of the Joint National Committee on Prevention, Detection, Evaluation, and Treatment of High Blood Pressure (JNC 7) for patients with diabetes mellitus or chronic kidney disease (CKD), the proportion of uncontrolled patients is even higher. Of NHANES participants with chronic kidney disease, only 37% were controlled to 130/80 mm Hg3, and only 25% of participants with diabetes were controlled to 130/85 mm Hg [1, 3].

An estimated 10–30% of hypertensive patients are resistant to treatment defined as uncontrolled blood pressure (BP) with the use of ≥3 medications, including a diuretic [4–10]. A large number of studies have demonstrated that patients with resistant hypertension compared with patients with controlled hypertension have significantly a higher rate of target organ damage; increased cardiovascular risk, including coronary heart disease, chronic kidney disease, congestive heart failure, and stroke; and a significantly poorer prognosis than those of nonresistant hypertensive patients [3, 11].

Poor medical adherence, poor blood pressure measuring technique, and white-coat effect are relevant challenges to figuring out the real burden of resistant hypertension [11].

Previous studies have shown that obesity is associated with resistant AH [12]. In addition, other studies have shown that diabetes is associated with a resistant hypertension [13, 14]. Studies show that resistant AH is associated with an increase in age, female gender, Negroid race, the presence of diabetes mellitus, obesity, chronic kidney disease, and left ventricular hypertrophy [1, 10, 15–19]. For early detection of resistant AH, aggressive therapy can reduce both cardiovascular morbidity and mortality. However, the exact prevalence of resistant AH is not known precisely because of its variety of definitions and diversity of study sites [20, 21].

Increased blood pressure is one of the most important risk factors for stroke [4, 6, 22], and uncontrolled hypertension increases this risk [1, 23]. The prevalence of hypertension in Asian countries is almost the same as in most developed countries; many Asian patients have uncontrolled hypertension compared with developed countries [24]. For example, in developed countries, blood pressure monitoring is about 52–60%, but in Malaysia, for example, this figure is 26% [2, 11, 24–26]. In Ukraine, blood pressure control in the urban population is 14% and in rural populations 8% [27, 28].

In the current study, we have assessed the prevalence of true resistant hypertension in a cohort of patients who take 3 and more antihypertensive agents, the clinical predictors of resistant hypertension.

2. Material and methods

The study included 1146 patients with resistant AH who received 3 or more antihypertensive drugs who were hospitalized in 2011–2015 at the Department of Symptomatic Hypertension at the Institute of Cardiology of Ukraine, Kyiv. The level of office blood pressure when admitted to the office when receiving 3 or more AH drugs was ≥140/90 mm Hg. The average systolic blood pressure (SBP)/diastolic blood pressure (DBP) was 174,60 ± 0,64/100,50 ± 0,38 mm Hg.

Inclusion criteria. The inclusion criteria are as follows: (1) men and women aged between 18 and 80 years old and (2) patients treated with 3 and more antihypertensive drugs. The diagnosis of RH was made after treatment with three antihypertensive drug classes at maximum tolerated doses for at least 6 months. The study did not include patients with acute myocardial infarction or cerebrovascular accidents less than 3 months, acute renal failure, decompensated liver disease (level of AST, ALT above 3 times upper limit of normal), pregnancy, or lactation.

Anthropometric measurements. Weight and height, measured by anthropometric scales, will be used to calculate the body mass index (BMI) using the formula

Parameters	Value
Men, n (%)	423 (36,91%)
Women, n (%)	723 (63,09%)
Height, m	1,7 ± 0,01
Weight, kg	87,7 ± 0,61
Age, years	57,9 ± 0,37
BMI, kg/m^2	31,0 ± 0,19
Office SBP mm Hg at hospitalization	174,6 ± 0,64
Office DBP mm Hg at hospitalization	100,5 ± 0,38
Office SBP mm Hg on discharge from hospital	131,3 ± 0,40
Office DBP mm Hg on discharge from hospital	80,1 ± 0,65
Index of apnea-hypopnea, events /h, (n = 75)	18,8 ± 2,11
AO, sm	3,4 ± 0,05
LA, sm	3,9 ± 0,06
Left ventricle end-systolic dimension, sm	3,5 ± 0,08
Left ventricle end-diastolic dimension, sm	5,5 ± 0,13
Left ventricle end-systolic volume, mL	48,3 ± 0,68
Left ventricle end-diastolic volume, mL	126,3 ± 1,17
Inter ventricular septum thickness, sm	1,2 ± 0,01
Left ventricle posterior wall thickness, sm	1,1 ± 0,01
EF, %	60,6 ± 0,23
Left ventricular mass index, g/m^2	138,2 ± 1,43
K, mmol/l	5,1 ± 0,49
Na, mmol/l	144,0 ± 0,30
Bilirubin, μmol/l	14,1 ± 0,21
Creatinine, μmol/l	87,2 ± 0,58
CKD-EPI, ml/min/1.73 m^2	82,7 ± 0,96
ALT, U/l	47,5 ± 3,39
AST, U/l	27,1 ± 0,71
Fasting glucose, mmol/l	6,0 ± 0,08
Uric acid, mmol/l (n = 850)	336,2 ± 3,62
Triiodothyronine (T3), microU/l (n = 94)	4,6 ± 1,16
Thyroxin (T4), microU/l (n = 111)	3,2 ± 0,67
Thyroid hormone (TTH), microU/l (n = 231)	2,2 ± 0,15
Metanephrine urine, microG/24 h (n = 95)	124,5 ± 7,10
Renin, ng/l (n = 89)	155,2 ± 79,73
Aldosterone, ng/l (n = 118)	29,4 ± 2,90
Aldosterone-renin ratio, c.u. (n = 75)	3,3 ± 0,68
Cortisol, ng/l (n = 21)	108,0 ± 30,73
Total cholesterol, mmol/l	5,5 ± 0,04
Triglycerides, mmol/l	1,6 ± 0,04
HDL cholesterol, mmol/l	1,3 ± 0,02
LDL cholesterol, mmol/l	3,3 ± 0,07
VLDL cholesterol, mmol/l	0,7 ± 0,02
IA (index of atherogenicity), c.u.	3,3 ± 0,07

Table 1.
Clinical and demographic characteristics of the examined patients (n = 1146).

BMI = weight (kg)/height squared (m^2). BMIs of 18.5–26.9 kg/m^2 are considered eutrophic values, while individuals with BMIs of 27.0–29.9 kg/m^2 are overweight and ≥30 kg/m^2 are obese.

All patients will undergo electrocardiography, echocardiography, and office-measured SBP and DBP; an oscillometric device will be used to calculate the average of the three measurements. Biochemical and imaging tests. Blood samples will be drawn from all patients at the first visits after fasting for 12 h to measure serum total cholesterol, high-density lipoprotein cholesterol (HDLc), low-density lipoprotein cholesterol (LDLc), very low-density lipoprotein cholesterol (VLDLc), triglycerides (TG), glucose, creatinine, sodium, and potassium; blood renin, aldosterone and aldosterone/renin ratio, metanephrines in urine, blood cortisol level, T3, T4, and TTH were determined in some patients. CKD-EPI was calculated. If needed, CT with intravenous contrast renal arteries and adrenal glands (for exclusion, secondary hypertension) and sleep apnea determination were performed on some patients (**Table 1**).

Statistical processing of the results was performed on a personal computer after creating databases in Microsoft Excel systems. The mean of the patients examined was determined using the analysis package in Microsoft Excel. All other statistical calculations were performed using SPSS 21.0. ANOVA to calculate the following parameters: the arithmetic mean value, M; the SD from the arithmetic mean value of m; and coefficient of reliability, p. The difference was considered reliable at a value of $p < 0.05$. The reliability of the difference between the groups was determined by the independent t-test for the mean. Correlation analysis was performed after determining the character of the distribution for Spearman.

3. Results

We examined 1146 patients who received 3 or more antihypertensive drugs. The mean age was 57,9 ± 0,37 years. The average body weight is 87,7 ± 0,61 kg. The average body mass index was 31,0 ± 0,19 kg/m^2. Average baseline office SBP and DBP were 174,6 ± 0,64 and 100,5 ± 0,38 mm Hg accordingly.

Most of the patients received 3 antihypertensive drugs: 51.4%. 48.6% of the patients take four to six drugs. Most of them take four drugs, 37.1%, 9.1% five drugs, and 2.4% six drugs.

The frequency of appointment of different classes of antihypertensive agents in the examined patients is presented in **Table 2**. In the structure of the appointments of antihypertensive drugs (AHD), ACE inhibitors were prescribed more often, 65.5%; calcium antagonists, 69.9%; and diuretics (loop, thiazide, and thiazide like), 91,8%. Beta-blockers were taken by 75.6% of patients and blockers to AT II receptor, 33.5%. Aldosterone receptor blockers were taken by 12.8% of patients and central activity drugs, 18.6%. Statins were taken by 63.8% of patients. Among those receiving combined therapy, most (71.9%) take fixed combinations. Attention was given to the low frequency of aldosterone receptor blockers received (12.8%), which was due to the fact that the study was conducted predominantly until the year 2015, when there was scientific evidence of the need for their use as the fourth drug.

For further analysis, we divided our patients according to the amount of drugs that they received. Although 3 or more antihypertensive drugs are used in the determination of resistant hypertension, we divided our patients into two groups: the first, those who took 3 drugs, and the second, those who took 4 or more drugs. A comparison of demographic characteristics and blood pressure levels in these groups is presented in **Table 3**.

Antihypertensive class	%
Angiotensin-converting enzyme inhibitors	65,5
Angiotensinogen receptor blockers	33,5
Calcium channel blockers	69,9
Diuretics (thiazide, loops, thiazide like)	91,8
β-Blockers	75,6
Mineralocorticoid receptor antagonists	12,8
Central-acting agonists	18,6
α-Blockers	2,6
Fixed combination	71,9
Acetylsalicylic acid	68,3
Statins	63,8

Table 2.
Frequency of appointment of antihypertensive drugs of different classes among patients with resistant arterial hypertension (n = 1146).

Parameters	Value		P
	3 drugs (n = 591)	**≥4 drugs (n = 555)**	
Men/women, n (%)	231 (39,0)/360 (60,8)	192 (34,6)/363 (65,4)	NS
Age, years	57,2 ± 0,54	58,7 ± 0,48	NS
Height, m	1,7 ± 0,01	1,7 ± 0,01	NS
Weight, kg	85,9 ± 0,85	89,3 ± 0,86	0,004
BMI, kg/m^2	30,2 ± 0,26	31,6 ± 0,27	<0,001
Office SBP mm Hg at hospitalization	170,4 ± 0,79	179,0 ± 1,00	<0,001
Office DBP mm Hg at hospitalization	98,4 ± 0,48	102,6 ± 0,58	<0,001
Office SBP mm Hg on discharge from hospital	129,3 ± 0,52	133,5 ± 0,59	<0,001
Office DBP mm Hg on discharge from hospital	78,5 ± 0,33	81,8 ± 1,29	0,013
Index of apnea-hypopnea, events/h[*]	18,7 ± 3,07	18,9 ± 2,90	NS

[*]*The somnography was performed in 31 patients in the first group and in 44 patients in the second group.*

Table 3.
Characteristics of patients, depending on the number of prescribed antihypertensive drugs (n = 1146).

As can be seen from the table, the groups did not differ by age and sex on average, but patients taking ≥4 drugs had significantly higher body mass and BMI than the group of patients taking 3 drugs. They also had significantly higher levels of office BP, both when they arrived in the hospital and on discharge. It should be noted that in both groups the value of the apnea-hypopnea index was high but did not differ from each other.

According to laboratory tests, patients taking ≥4 drugs had significantly higher blood glucose levels—(6.20 ± 0.09) mmol—than patients taking 3 drugs, (5.90 ± 0.14) mmol (p < 0.05), and a higher level of renin plasma ((218.30 ± 15.73) vs. (31.10 ± 5.91) ng/l, (p < 0.05)). Renin-aldosterone ratio was almost twice as high

(3.90 ± 0.95 vs. 2.20 ± 0.65 U/d), although it did not differ significantly between patient groups.

Analyzing the differences between patients in both groups in the main clinical states, we found that in patients receiving 4 or more drugs, significantly more obesity (45.6 vs. 34.0%), more often secondary hypertension (5.4 against 2.5%) due to stenosis of the renal arteries (1.6 vs. 0.2%), and hyperaldosteronism (2.3 vs. 0.2%) were observed; type II diabetes mellitus was more frequent (24.7 vs. 9, 5%); pathology of the thyroid gland (12.6 vs. 8.4%) due to hypothyroidism (4.1 vs. 1.5%) and chronic kidney disease were more common (5.0 vs. 1.5%); and chronic pyelonephritis (18.9 vs. 14.0%), ischemic heart disease (IHD) (47.0 vs. 37.2%) with angina pectoris III (7.7 vs. 3.4%), and heart failure (15.0 vs. 9.3%) were observed.

In analyzing the degree of decrease in blood pressure, we found that, in general, the reduction of office blood pressure among patients receiving 3 or more drugs was for SBP (43,47 ± 0,65) mm Hg and for DBP (20,33 ± 0.74) mm Hg, $p < 0.001$ for both values.

The analysis of the degree of reduction of blood pressure, depending on the amount of drugs, showed that DBP between patients taking 3 and 4 or more drugs did not differ significantly, 19.88 versus 20.81 mm Hg, respectively, and office SBP significantly lowered in patients taking 4 or more drugs at 45.78 mm Hg vs. group taking 3 drugs—41.3 mm Hg, $p < 0.001$.

Among all patients (n = 1146), 355 (31%) did not reach the target blood pressure level. Patients who did not achieve targeted SBP (31%) had significantly higher blood pressure when inpatient. They had a significantly higher blood cortisol

	SBP on discharge		DBP on discharge	
Age	β = −0,089	P < 0,001	β = −0,196	P < 0,001
Gender	β = 0,125	P < 0,001	β = 0,130	P < 0,001
Weight	β = 0,106	P = 0,004	β = 0,127	P < 0,001
BMI			β = −0,202	P = 0,010
Arterial hypertension	β = 0,205	P = 0,001	β = 0,117	P < 0,001
Arterial hypertension II	β = 0,108	P = 0,001	β = 0,133	P < 0,001
Arterial hypertension III	β = −0,259	P = 0,023	β = −0,106	P = 0,001
Secondary arterial hypertension	β = −0,096	P = 0,006	β = −0,114	P < 0,001
Vasorenal arterial hypertension	β = −0,082	P = 0,011	β = −0,101	P = 0,002
Hyperaldosteronism	β = −0,103	P = 0,022	β = −0,071	P = 0,027
Pituitary adenoma			β = −0,089	P = 0,006
Heart failure			β = 0,146	P < 0,001
Heart failure II	β = −0,079	P = 0,014	β = −0,103	P = 0,002
Adrenal pathology	β = −0,065	P = 0,043	β = −0,070	P = 0,029
T4	β = −0,230	P = 0,020	β = −0,219	P = 0,027
Interventricular septum thickness	β = 0,214	P = 0,027	β = 0,154	P < 0,001
HD cholesterol	β = −0,230	P = 0,001	β = −0,198	P = 0,003
Calcium channel blockers	β = −0,140	P < 0,001	β = −0,080	P = 0,013
Central-acting agonists	β = −0,214	P < 0,001	β = −0,137	P < 0,001
Mineralocorticoid receptor antagonists	β = −0,101	P = 0,002	β = −0,096	P = 0,005
Acetylsalicylic acid	β = 0,169	P < 0,001	β = 0,197	P < 0,001

Table 4.
Factors influencing the reduction of blood pressure on discharge from the hospital (n = 1146).

level (155.0 ± 44.0 vs. 35.9 ± 20.8 ng/l) and the highest left ventricular mass index (147.5 ± 3.46 vs. 135.3 ± 1.74 g/m^2), and obesity (42.9 vs. 37.5%), kidney abnormality (2.7 vs. 0.8%), obliterative lower extremity atherosclerosis (2.0 vs. 0.2%), structural alterations in the adrenal gland (3.0 vs. 1.2%), nephropathy (1.3 vs. 0.2%), and higher heart failure (HF in the 16.9 vs. 8.5%) were more common.

When we performed a regression analysis, we found that the decrease in office systolic and diastolic blood pressure depended on age, sex, and body weight. Thus, office SBP/DBP was worse in men; patients with a younger age; patients with greater body mass, with hypertension III degree, and with secondary hypertension, especially with hyperaldosteronism and vasorenal hypertension, adrenal pathology, heart failure, lower T4 hormone levels, more low levels of HDL cholesterol, and a larger thickness of interventricular septum thickness; and patients receiving less calcium antagonists, centrally acting drugs, and aldosterone antagonists. Data are presented in **Table 4**.

Table 5 presents the results of a regression analysis for the detection of predictors of resistance hypertension in patients receiving 3 or more antihypertensive drugs. The predictors for blood pressure reduction were male sex, left ventricular mass index, interventricular septum thickness, left ventricle posterior wall thickness, hypothyroidism, the presence of chronic kidney disease, CKD-EPI level, blood creatinine levels, the presence of heart failure, the presence of secondary hypertension, hyperaldosteronism, pituitary adenoma, and vasorenal hypertension.

Variables	Predictors of failure to reach the target BP	
Gender	β = 0,119	P < 0,001
Left ventricular mass index	β = 0,139	P = 0,001
Inter ventricular septum thickness	β = 0,169	P = 0,002
Left ventricle posterior wall thickness	β = 0,147	P < 0,001
Arterial hypertension	β = 0,085	P = 0,009
Arterial hypertension II	β = 0,090	P = 0,005
Arterial hypertension III	β = 0,077	P = 0,018
Secondary arterial hypertension	β = 0,075	P = 0,020
Vasorenal arterial hypertension	β = 0,107	P = 0,001
Hyperaldosteronism	β = 0,064	P = 0,049
Pituitary adenoma	β = 0,068	P = 0,036
Pathology of the thyroid gland	β = 0,102	P = 0,002
Hypothyroidism	β = 0,069	P = 0,031
Mixed goiter	β = 0,072	P = 0,024
Heart failure II	β = 0,125	P < 0,001
Chronic kidney disease	β = 0,076	P = 0,018
Creatinine	β = 0,108	P = 0,028
CKD-EPI	β = 0,135	P = 0,025
Office SBP mm Hg at hospitalization	β = 0,368	P < 0,001
Office DBP mm Hg at hospitalization	β = 0,238	P < 0,001
Calcium channel blockers	β = −0,116	P = 0,001
Central-acting agonists	β = −0,146	P < 0,001
Fixed combinations	β = 0,098	P = 0,003
Acetylsalicylic acid	β = 0,103	P = 0,002

Table 5.
Predictors of resistance hypertension in patients receiving 3 or more antihypertensive drugs (n = 1146).

4. Discussion

Our data showed that in patients who received 3 or more antihypertensive drugs in 31%, the goal blood pressure (<140/90 mm Hg) was not reached, meaning it was true resistant hypertension. This is possible somewhat more than in other studies, but we have a specialized department, which is directed precisely by patients who failed to reach the target levels of blood pressure at the outpatient stage.

Among our patients with resistant arterial hypertension, 3 antihypertensive drugs were received by 51.4% of patients, 4 antihypertensive drugs were taken by 37.1% of patients, 5 antihypertensive drugs were taken by 9.1% of patients, and 6 antihypertensive drugs were taken by 2.4% of patients.

In our study ACE inhibitors were more often prescribed in 65.5% of patients, calcium antagonists in 69.9% of patients, and diuretics in 91.8% of patients. Beta-adrenergic blockers were administered to 75.5% of patients, receptor blockers to AT II to 33.5% of patients, and aldosterone receptor blockers to 12.8% of patients. Fixed combinations were taken by 71.9% of patients.

In our study patients who did not achieve targeted SBP (31%) had significantly higher blood pressure when inpatient. They had a significantly higher blood cortisol level (155.0 ± 44.0 vs. 35.9 ± 20.8 ng/l) and the highest left ventricular mass index (147.5 ± 3.46 vs. 135.3 ± 1.74 g/m^2), and obesity (42.9 vs. 37.5%), kidney abnormality (2.7 vs. 0.8%), obliterative lower extremity atherosclerosis (2.0 vs. 0.2%), structural alterations in the adrenal gland (3.0 vs. 1.2%), nephropathy (1.3 vs. 0.2%), and more often heart failure (16.9 vs. 8.5%) were more common.

In our study, patients taking ≥ 4 drugs had significantly higher blood glucose levels—(6.20 ± 0.09) mmol/l—than patients taking 3 drugs, (5.90 ± 0.14) mmol/l ($p < 0.05$), and the highest level of renin plasma ((218.30 ± 15.73) vs. (31.10 ± 5.91) ng/l ($p < 0.05$)). Renin-aldosterone ratio was almost twice as high (3.90 ± 0.95 vs. 2.20 ± 0.65 U/d), although it did not differ significantly between patient groups.

Yook Chin Chia and Siew Mooi Ching studied the prevalence and predictors of resistance to hypertension in Southeast Asia [24]. The prevalence of resistant hypertension in the primary examination in their study was 8.8%. Their data also show that patients with chronic kidney disease were 2.9-fold more likely to develop resistant AH than in patients without CKD. This is consistent with the findings in other studies [25, 29]. In our study, CKD-EPI and CKD were predictors of resistance hypertension.

The authors explain this by the fact that in patients with CKD, there is increased sensitivity to salt, resulting in a delay in sodium and fluid, which leads to more complex control of blood pressure [24]. Patients in the Yook Chin Chia study were aged 66.9 years. In our study, patients were more younger, 57.9 ± 0.37. Resistant hypertension in Yook Chin Chia study was negatively related to age, which the authors explain with the effect of survival of patients who were treated compared with those patients who already had complications from uncontrolled hypertension. In our study, office SBP/DBP was worse in younger age patients.

Many studies have shown that most patients with hypertension need 2 or more drugs to achieve the target blood pressure [2, 30–34]. The average number of AH drugs used in their study was 2. They also showed poor monitoring of blood pressure among those taking only 2 drugs; even in those who received 3 drugs, the level of blood pressure control was less than 50%. However, in general, the use of diuretics in their study was low.

Holmqvist et al. studied the adherence to treatment in patients with resistance to hypertension, which controlled or did not control blood pressure and what factors contributed to nonadherence to treatment. 5846 patients received treatment with 3 or more AH drugs for 2 years [26]. Patients who achieved target blood pressure levels were older in age and among them those with diabetes were fewer. Initially, patients had an adherence above 80%. During the first year of treatment, the adherence decreased by 11%, regardless of whether it was controlled or not controlled. The highest adherence was observed only in patients with diabetes mellitus and hypertension, in which the authors explain the structuring of the treatment of such a patient.

Resistant hypertension is associated with significant adverse effects, including an increased risk of cardiovascular events and death as well as a decrease in the quality of life [22, 24]. The exact mechanisms underlying the development of resistant hypertension remain unclear, although several mechanisms were proposed [8, 9, 35, 36]. An increase in fluid content and an increase in the level of aldosterone play a crucial role in the development of resistance hypertension, whereas enhanced activation of the sympathoadrenal system significantly contributes to refractory hypertension [14, 20, 34].

In conclusion, the predictors for blood pressure reduction were male sex, left ventricular mass index, interventricular septum thickness, left ventricle posterior wall thickness, the presence of chronic kidney disease, CKD-EPI level, blood creatinine levels, and the presence of heart failure.

Disclosure

None.

Author details

Yuriy Mykolayovych Sirenko, Oksana Leonidivna Rekovets and
Olena Oleksandrivna Torbas*
FI "NSC" Institute of Cardiology named after M.D. Strazhesko NAMS of Ukraine,
Kyiv, Ukraine

*Address all correspondence to: olenatorbas@gmail.com

IntechOpen

References

[1] Hajjar I, Kotchen TA. Trends in prevalence, awareness, treatment, and control of hypertension in the United States, 1988-2000. Journal of the American Medical Association. 2003;**290**:199-206

[2] Lloyd-Jones DM, Evans JC, Larson MG, O'Donnell CJ, Rocella EJ, Levy D. Differential control of systolic and diastolic blood pressure: Factors associated with lack of blood pressure control in the community. Hypertension. 2000;**36**:594-599

[3] Chobanian AV, Bakris GL, Black HR, et al. The seventh report of the Joint National Committee on prevention, detection, evaluation, and treatment of high blood pressure: The JNC 7 report. Journal of the American Medical Association. 2003;**289**(19):2560-2572

[4] Hwang AY, Dietrich E, Pepine CJ, Smith SM. Resistant hypertension: Mechanisms and treatment. Current Hypertension Reports. 2017;**19**:56. DOI: 10.1007/s11906-017-0754-x

[5] Bangalore S, Fayyad R, Laskey R, et al. Prevalence, predictors, and outcomes in treatment-resistant hypertension in patients with coronary disease. The American Journal of Medicine. 2014;**127**(1):71-81.e71

[6] Benjamin EJ, Blaha MJ, Chiuve SE, et al. Heart disease and stroke statistics—2017 update: A report from the American Heart Association. Circulation. 2017;**135**:e146-e603

[7] Brandani L. Resistant hypertension: A therapeutic challenge. Journal of Clinical Hypertension. 2018;**20**:76-78

[8] Roush GC, Holford TR, Guddati AK. Chlorthalidone compared with hydrochlorothiazide in reducing cardiovascular events: Systematic review and network meta-analyses. Hypertension. 2012;**59**(6):1110-1117

[9] Sarafidis PA, Georgianos P, Bakris GL. Resistant hypertension—Its identification and epidemiology. Nature Reviews Nephrology. 2013;**9**(1):51-58

[10] Williams B, Mancia G, Spiering W, et al. ESC/ESH guidelines for the management of arterial hypertension the task force for the management of arterial hypertension of the European Society of Cardiology (ESC) and the European Society of Hypertension (ESH). European Heart Journal. 2018;**00**:1-98. DOI: 10.1093/eurheartj/ehy339

[11] Carris NW, Ghushchyan V, Libby AM, Smith SM. Health-related quality of life in persons with apparent treatment-resistant hyper- tension on at least four antihypertensives. Journal of Human Hypertension. 2016;**30**(3):191-196

[12] Hall JE, do Carmo JM, da Silva AA, Wang Z, Hall ME. Obesity-induced hypertension: Interaction of neurohumoral and renal mechanisms. Circulation Research. 2015;**116**(6):991-1006

[13] Hwang AY, Dave C, Smith SM. Trends in antihypertensive medication use among US patients with resistant hypertension, 2008 to 2014. Hypertension. 2016;**68**(6):1349-1354

[14] Pimenta E, Gaddam KK, Oparil S, et al. Effects of dietary sodium reduction on blood pressure in subjects with resistant hypertension: Results from a randomized trial. Hypertension. 2009;**54**(3):475-481

[15] Garg JP, Elliott WJ, Folker A, Izhar M, Black HR. Resistant hypertension revisited: A comparison of two university-based cohorts.

American Journal of Hypertension. 2005;**18**(5 Pt 1):619-626

[16] Gerritsen J, Dekker JM, TenVoorde BJ, et al. Impaired autonomic function is associated with increased mortality, especially in subjects with diabetes, hypertension, or a history of cardiovascular disease: The Hoorn study. Diabetes Care. 2001;**24**(10):1793-1798

[17] Gifford RW Jr, Tarazi RC. Resistant hypertension: Diagnosis and management. Annals of Internal Medicine. 1978;**88**(5):661-665

[18] Irvin MR, Booth JN 3rd, Shimbo D, et al. Apparent treatment- resistant hypertension and risk for stroke, coronary heart disease, and all-cause mortality. Journal of the American Society of Hypertension. 2014;**8**(6):405-413

[19] Le Jemtel TH, Richardson W, Samson R, Jaiswal A, Oparil S. Pathophysiology and potential non-pharmacologic treatments of obesity or kidney disease associated refractory hypertension. Current Hypertension Reports. 2017;**19**(2):18

[20] Taler SJ, Textor SC, Augustine JE. Resistant hypertension: Comparing hemodynamic management to specialist care. Hypertension. 2002;**39**(5):982-988

[21] Vlase HL, Panagopoulos G, Michelis MF. Effectiveness of furosemide in uncontrolled hypertension in the elderly: Role of renin profiling. American Journal of Hypertension. 2003;**16**(3):187-193

[22] Siddiqui M, Dudenbostel T, Calhoun DA. Resistant and refractory hypertension: Antihypertensive treatment resistance vs treatment failure. The Canadian Journal of Cardiology. 2016;**32**(5):603-606

[23] Hermida RC, Ayala DE, Smolensky MH, Fernandez JR, Mojon A,

Portaluppi F. Chronotherapy with conventional blood pressure medications improves management of hypertension and reduces cardiovascular and stroke risks. Hypertension Research. 2016;**39**(5):277-292

[24] Chia YC, Ching SM. Prevalence and predictors of resistant hypertension in a primary care setting: A cross-sectional study. BMC Family Practice. 2014;**15**:131. Available from: http://www.biomedcentral.com/1471-2296/15/131

[25] Feng W, Dell'Italia LJ, Sanders PW. Novel paradigms of salt and hypertension. Journal of the American Society of Nephrology. 2017;**28**(5):1362-1369

[26] Holmqvist L, Boström KB, Kahan T, Schiöler L, Qvarnström M, Wettermark B, et al. Drug adherence in treatment resistant and in controlled hypertension-results from the Swedish primary care cardiovascular database (SPCCD). Pharmacoepidemiology and Drug Safety. 2018;**27**(3):315-321. DOI: 10.1002/pds.4388

[27] Gorbas IM, Smirnova IP, Vakalyuk IP, et al. Epidemiolohichna sytuatsia arterialnoy hypertensii u silskiy populyatsii in Ukraine. Liku Ukraine. No. 7. 2013. pp. C88-C91

[28] Kvasha EA, Gorbas IM, Smirnova IP, Sribna IV. Arterialna hypertensia: 35-letnia dinamika rasprostranionosti i effektivnosti kontrolya na populyatsionnom urovne sredi muzchin v. No. 3. 2016. pp. C18-C23

[29] Duprez DA. Aldosterone and the vasculature: Mechanisms mediating resistant hypertension. Journal of Clinical Hypertension. 2007;**9**(1 Suppl 1):13-18

[30] Ernst ME, Moser M. Use of diuretics in patients with hypertension. The

New England Journal of Medicine. 2009;**361**(22):2153-2164

[31] Gaddam KK, Nishizaka MK, Pratt-Ubunama MN, et al. Characterization of resistant hypertension: Association between resistant hypertension, aldosterone, and persistent intravascular volume expansion. Archives of Internal Medicine. 2008;**168**(11):1159-1164

[32] Humphrey JD, Harrison DG, Figueroa CA, Lacolley P, Laurent S. Central artery stiffness in hypertension and aging: A problem with cause and consequence. Circulation Research. 2016;**118**(3):379-381

[33] Kumbhani DJ, Steg PG, Cannon CP, et al. Resistant hypertension: A frequent and ominous finding among hypertensive patients with atherothrombosis. European Heart Journal. 2013;**34**(16):1204-1214

[34] Mills KT, Bundy JD, Kelly TN, et al. Global disparities of hypertension prevalence and control: A systematic analysis of population-based studies from 90 countries. Circulation. 2016;**134**(6):441-450

[35] Rimoldi SF, Scherrer U, Messerli FH. Secondary arterial hypertension: When, who, and how to screen? European Heart Journal. 2014;**35**(19):1245-1254

[36] Zygmuntowicz M, Owczarek A, Elibol A, Olszanecka- Glinianowicz M, Chudek J. Blood pressure for optimal health- related quality of life in hypertensive patients. Journal of Hypertension. 2013;**31**(4):830-839

www.ingramcontent.com/pod-product-compliance
Lightning Source LLC
Chambersburg PA
CBHW081239190326
41458CB00016B/5841